SpringerBriefs in Mathematics

SpringerBriefs in Mathematics showcase expositions in all areas of mathematics and applied mathematics. Manuscripts presenting new results or a single new result in a classical field, new field, or an emerging topic, applications, or bridges between new results and already published works, are encouraged. The series is intended for mathematicians and applied mathematicians.

More information about this series at http://www.springer.com/series/10030

SBMAC SpringerBriefs

The **SBMAC SpringerBriefs** series publishes relevant contributions in the fields of applied and computational mathematics, mathematics, scientific computing and related areas. Featuring compact volumes of 50 to 125 pages, the series covers a range of content from professional to academic.

The Sociedade Brasileira de Matemática Aplicada e Computacional (Brazilian Society of Computational and Applied Mathematics, SBMAC) is a professional association focused on computational and industrial applied mathematics. The society is active in furthering the development of mathematics and its applications in scientific, technological and industrial fields. The Brazilian Society of Applied and Computational Mathematics has helped in developing the applications of mathematics in science, technology and industry, in encouraging the development and implementation of effective methods and mathematical techniques to be applied for the benefit of science and technology and in promoting the exchange of ideas and information between the areas of mathematical applications.

http://www.sbmac.org.br/   Sociedade Brasileira de Matemática Aplicada e Computacional

Luciana Takata Gomes
Laécio Carvalho de Barros • Barnabas Bede

Fuzzy Differential Equations in Various Approaches

 Springer

Luciana Takata Gomes
Department of Chemistry,
  Physics and Mathematics
Federal University of São Carlos - Campus
  Sorocaba
Sorocaba, São Paulo, Brazil

Laécio Carvalho de Barros
Institute of Mathematics, Statistics
  and Scientific Computation
Department of Applied Mathematics
University of Campinas
Campinas, São Paulo, Brazil

Barnabas Bede
Department of Mathematics
DigiPen Institute of Technology
Redmond Washington, USA

ISSN 2191-8198   ISSN 2191-8201   (electronic)
SpringerBriefs in Mathematics
ISBN 978-3-319-22574-6   ISBN 978-3-319-22575-3   (eBook)
DOI 10.1007/978-3-319-22575-3

Library of Congress Control Number: 2015947497

Mathematics Subject Classification (2010): 34A07, 26E50, 47S40

Springer Cham Heidelberg New York Dordrecht London

Printed on acid-free paper

Springer International Publishing AG Switzerland is part of Springer Science+Business Media (www.
springer.com)

To my family, friends, and Tiago
To my parents, brother, and sisters
To my family

Preface

Differential equations have been widely explored in many fields, from applications in physics, engineering, economics, and biology to theoretical mathematical developments. Its presence in undergraduate and graduate courses of the aforementioned areas and countless textbooks and papers ensures their usefulness and importance.

A much newer theory, fuzzy sets theory, created to model subjective concepts whose boundaries are nonsharp, has also been explored in various fields due to its great applicability and functionality. As soon as the idea of a function with fuzzy values was born, it raised the idea of some kind of fuzzy differential equation (FDE) as well. Since then, researches defined different fuzzy derivatives and fuzzy functions, giving rise to different theories of FDEs. Great part of its development is in papers and rare textbooks, which usually dedicate few sections to the subject. An updated textbook entirely devoted to FDEs has been missing and this book is intended to cover this gap.

This book is aimed at researchers and graduate students interested in FDEs. It may be useful to scientists of areas such as engineering, biology, and economics dealing with uncertain dynamical systems and fuzzy concepts, besides mathematicians interested in theoretical developments. The text focuses on fuzzy initial value problems (FIVPs) and is intended to be a reference textbook with the basics of various approaches of FDEs. The best known approaches—via Hukuhara derivative, fuzzy differential inclusions (FDIs) and via extension of the solution are presented, as well as the recent strongly generalized derivative and the extension of the derivative operator. This book is the result of years of study aimed at (but not restricted to) developing the last approach, including new results related to it.

The theory of FDEs via the extension of the derivative operator is based on fuzzy calculus for fuzzy bunches of functions. This kind of function is a departure from the generally known fuzzy-set-valued functions. A deeper understanding of the different kinds of fuzzy functions is needed, which we endeavor to offer to the reader. Comparisons and links among all the mentioned approaches of FDEs are provided through an original interpretation that situates the novel theory of the extension of the derivative operator as the missing link needed to fill the gap to connecting all approaches.

The reader is not required to be conversant in fuzzy sets, though it is desired; the book is intended to cover all the necessary prerequisites in this subject. In order to understand all the demonstrations, the reader must know basic functional analysis, but it is not mandatory in order to understand the theory as a whole.

Some highlights that make this book unique are summarized next:

- The text presents the most known approaches of FDEs in an unprecedented view and with comprehensive historical overview on the subject.
- The book scrutinizes the recent theory of FDEs via extension of the derivative operator, presenting it for the first time in a textbook.
- The reader is not expected to be conversant in fuzzy sets. A chapter with basic concepts and illustrative examples is dedicated to eliminate deficiencies.
- The text presents theoretical depth, though it is also intended to serve as a useful text to researches from application areas.

The authors would like to express their gratitude to Professor Geraldo Silva and the Springer staff for the assistance provided in preparing the manuscript. The first and second authors acknowledge CNPq of the Ministry for Science and Technology of Brazil for financial support.

Sorocaba, Brazil   Luciana Takata Gomes
Campinas, Brazil   Laécio Carvalho de Barros
Redmond, WA, USA   Barnabas Bede

Contents

Symbols

a, f, x	(Lowercase) Non-fuzzy elements
A, F, X	(Capital letters) Fuzzy or non-fuzzy subsets
$\mathbb{U}, \mathbb{V}$	General non-fuzzy universes
$\mathcal{K}^n$	The family of all nonempty compact subsets of $\mathbb{R}^n$
$\mathcal{K}_{\mathscr{C}}^n$	The family of all nonempty compact and convex subsets of $\mathbb{R}^n$
$\mathscr{P}(\mathbb{U})$	Powerset of $\mathbb{U}$
$\mathscr{F}(\mathbb{U})$	The family of all fuzzy subsets of $\mathbb{U}$
$\mathscr{F}_{\mathcal{K}}(\mathbb{U})$	The family of all fuzzy subsets of $\mathbb{U}$ with nonempty compact α-cuts
$\mathscr{F}_{\mathscr{C}}(\mathbb{U})$	The family of all fuzzy subsets of $\mathbb{U}$ with nonempty compact and convex α-cuts
$[A]_\alpha$	α-cut of the fuzzy subset A
a_-^α, a_+^α	Lower and upper endpoints of the α-cut of the fuzzy number A, that is, $[A]_\alpha = [a_-^\alpha, a_+^\alpha]$
$\hat{f}$	Extension of function f (see Sect. 2.2)
$\hat{D}$	Extension of the derivative operator D (see Sect. 3.2.2)
$\hat{\int}$	Extension of the integral operator $\int$ (see Sect. 3.2.1)
$a \wedge b$	Minimum of a and b: $a \wedge b = \min\{a, b\}$
d_∞	Pompeiu–Hausdorff metric for spaces of fuzzy subsets (see Sect. 2.4)
d_E	Endographic metric for spaces of fuzzy subsets (see Sect. 2.4)
d_p	L^p type metric for spaces of fuzzy subsets (see Sect. 2.4)
$\overline{B}(X, q)$	Closed ball $\overline{B}(X, q) = \{A \in \mathscr{F}_{\mathscr{C}}(\mathbb{U}) : d_\infty(X, A) \leq q\}$
$F_H'(x)$	Hukuhara derivative of the fuzzy function F at x (see Sect. 3.1.2)
$F_G'(x)$	Strongly generalized derivative of the fuzzy function F at x (see Sect. 3.1.2)
$F_{gH}'(x)$	Generalized Hukuhara derivative of the fuzzy function F at x (see Sect. 3.1.2)
$F_g'(x)$	Fuzzy generalized derivative of the fuzzy function F at x (see Sect. 3.1.2)

$E(I; \mathbb{U})$ General space of functions from the interval I to $\mathbb{U}$

$C(I; \mathbb{R}^n)$ Space of continuous functions from I to $\mathbb{R}^n$

$\mathscr{A}C(I; \mathbb{R}^n)$ Space of absolutely continuous functions from I to $\mathbb{R}^n$ (see Appendix A)

$L^p(I; \mathbb{R}^n)$ L^p space of functions from I to $\mathbb{R}^n$ (see Appendix A)

Chapter 1
Introduction

Fuzzy systems were created to overcome the binary reasoning deep-rooted in the classical logic and mathematics. Under the dichotomous thinking, statements are completely true or completely false and elements are totally in or totally out of a set—never in the halfway or in different degrees. The creation of fuzzy systems allowed for reproducing the human reasoning in a computer understandable language and became successful when applied to modeling engineering problems. The modeling of any phenomena by humans is subjected to the limitation of the human being in understanding, collecting data, interpreting, and concluding, aside from their subjective reasoning. Moreover, the classification of the most various objects is subjected to the possible nonsharp boundaries inherent to the definition of classes made by humans. For instance, there is no exact bound in the definition of the group of the "populous cities." Taking one person by one out of a populous city we can reduce its population to zero without ever experiencing the exact moment when we think "now if I take one individual out this city is no more populous."

Nowadays fuzzy sets and fuzzy logic are present in various fields, from applications in the industry passing through applications in natural phenomena and psychology to mathematical theoretical aspects. What applied areas have in common is the presence of vague and uncertain information and the modeling done by the human being, whose reasoning is subjective, imprecise and, not uncommonly, even contradictory. Since mathematical tools are used to modeling all these applications, its theoretical aspects have to admit the concept of "fuzziness," that is, the partial truth of a statement or the partial membership of an element to a subset. Mathematical concepts such as numbers, sets, metrics, functions, operators, now have their "fuzzy versions."

Fuzzy Set Theory Is Not Fuzzy One should not think, however, that the fuzziness admitted in a natural phenomena and the tools created to deal with it make the theory fuzzy itself. Zimmermann in [40] clarifies:

© The Author(s) 2015
L.T. Gomes et al., *Fuzzy Differential Equations in Various Approaches*,
SpringerBriefs in Mathematics, DOI 10.1007/978-3-319-22575-3_1

Fuzzy set theory provides a strict mathematical framework (there is nothing fuzzy about fuzzy set theory!) in which vague conceptual phenomena can be precisely and rigorously studied.

In truth, taking into account the vague information is more realistic than presuming it crisp and precise, if it is not in fact.

Fuzzy Differential Equations Modeling of various phenomena frequently makes use of differential equations. In order to include imprecision, the fuzzy approach is often used. In particular, differential inclusions and, more recently, FDEs, or even fuzzy differential inclusions (FDIs) have been used.

In population dynamics, for instance, [23] recalls that individuals may exhibit some preferences or strategies, that is, inside a group they do not behave all in the same manner. Environmental or demographic noise is also a source of uncertainty. Via standard theory of differential equations, it is not possible to take these factors into account.

The authors of [23, 24] claim for the use of FDIs in population dynamics. According to these studies, the stochastic approach, via the use of white noise (a linear term in the differential equation) to model the uncertainty of the dynamics, is not the most appropriate one. The probabilistic approach would be suitable for the "hard sciences," such as physics and electronics, not for a "soft science" as biology. The white noise would emphasize short time scales and would lead to mathematically tractable models, hence it was used to treat many problems, but there are many others that demand for a different approach. The alternative would be the deterministic noise, including what they call the unknown-but-bounded-noise, i.e., the imprecision enters the dynamics via a parameter whose only assumption is that its values belong to a bounded set U, which may depend on time or the state variable. This approach leads to differential inclusions (see [2]) and considering some kind of "preference" of some parameter(s) in U determines a higher membership degree of a more suitable solution. This characterizes FDIs.

It is also possible to define fuzzy derivatives and consider the function as fuzzy as it has been done by many authors [5, 8, 13, 19, 33, 35, 36]. The first proposal, based on the Hukuhara derivative for interval-valued functions, has been criticized for presenting nondecreasing fuzziness. In other words, a dynamical system whose initial uncertainty is different from zero does not evolve to non-uncertain states. In the fuzzy context, it means that the solution cannot reach a nonfuzzy value. This situation is not consistent with the nuclear decay model, for instance. Other cited approaches succeeded to overcome this shortcoming and each of them is based on different notions of functions, differentiability, solutions, such that the solutions to FIVPs may differ greatly from each other.

1.1 Initial Value Problems

This book treats fuzzy initial value problems (FIVP). An initial value problem (IVP) is a system of an ordinary differential equation (ODE) together with a value called initial condition:

$$\begin{cases} x'(t) = f(t, x(t)) \\ x(0) = x_0 \end{cases}. \tag{1.1}$$

In what we call the "classical case," the solution is usually defined as a real-valued continuous function $x(\cdot)$ that satisfies the initial condition and the differential equation at every t in a given domain. The symbol $x'(t)$ stands for the derivative of x at t. The function $x(\cdot)$ is interpreted as a curve such that the velocity and the direction to be followed are determined by the function f, at each real value t. In this case, the solution is a real-valued function with real-valued argument. When the context demands distinction from the fuzzy case, we will call the IVP a "classical IVP" and the solution a "classical solution."

This approach is widely used to model physical, biological, chemical phenomena. In a biological interpretation, x can be the number of individuals of a population (ants, fishes, predators, humans, viruses, infected people), t is time, and f is the *rate* with which the population changes in quantity. The ability of modeling various phenomena, as well as theorems regarding existence of solution and practical techniques to find it (analytically or numerically), justifies wide use of IVPs.

Fuzzy set theory treats of sets in universes such that the elements have partial membership degree. That is, it is admissible that an element is not completely in or completely out of the set, but presents an intermediate degree. The success of fuzzy set theory, specially in modeling some control problems, has generated interest in many fields. Several concepts of the "nonfuzzy" theory were extended to the fuzzy case. This is no different in differential equations theory.

1.2 Fuzzy Initial Value Problem

IVP (1.1) becomes a new problem if any parameter presents fuzziness and it is called FIVP. The first time the term *fuzzy differential equation* (*FDE*) was used was in [21] and only in 1987 did FDE take on characteristic of the way it is used nowadays [19, 35]. Reference [19] made use of the Hukuhara derivative for fuzzy-set-valued functions and [35] used an equivalent definition. In both studies, the FIVP was defined using an FDE and a fuzzy initial value:

$$\begin{cases} X'(t) = F(t, X(t)) \\ X(0) = X_0 \end{cases}. \tag{1.2}$$

The function F is a fuzzy-set-valued function, that is, its values are fuzzy sets. Hence, the derivative X' of the unknown function X is also fuzzy. It means that the direction to be followed by the solution is a fuzzy set and the solution, at each t, is a fuzzy set as well.

There is also an integral equation associated with (1.2) as in the classical case [19]. It involves a fuzzy integral and the Minkowski sum and one realizes that the solution to this kind of equation always has nondecreasing diameter. That is, the *fuzziness* (or, according to the interpretation, the *uncertainty*) does not decrease with time. As it will be fully explained in Sect. 4.2, this is considered a shortcoming since this is not expected from phenomena such as decay in population dynamics.

Generalizations of the Hukuhara derivative fixed this defect (see next Sects. 1.2.1 and 4.3), but before that, other interesting approaches emerged and are still being intensively studied, namely the FDIs and extension of the solution. These two approaches are based on a completely different view of fuzzy FDEs. Though they receive this classification, they are not really FDEs. There is no equality between the derivative of a fuzzy function and the function that determines the direction of the dynamic since there is no derivative of fuzzy function. The derivative is that of classical functions. In the FDIs case, at each pair (t, x), there are different possible values of the function f, each one with a membership degree to the set "fuzzy direction field." The membership degree of the initial solution (to the set "fuzzy initial condition") and the direction field establish the membership degree of a nonfuzzy function (or its attainable set) to the solution of the FIVP. The common approach of extension of solution solves classical differential equations. The initial condition and fuzzy parameters determine the solution, which is usually a fuzzy-set-valued function.

A novel idea has been recently developed and connects both mentioned interpretations for FIVPs [5, 16, 17]. A fuzzy derivative is proposed, defined via extension of the derivative operator, denoted by $\hat{D}$, and it turns out to be based on differentiating classical functions. Moreover, an FDE has to be satisfied. Comparisons of the results between this and the other approaches are inevitable and, in fact, the new derivative leads to the same solutions produced by the other methods, provided some conditions are satisfied (see Sect. 3.3).

In summary:

- A solution to IVP (1.1) is a function $x(\cdot) \in E([a, b]; \mathbb{R}^n)$, where $E([a, b]; \mathbb{R}^n)$ is a space of functions from $[a, b]$ to $\mathbb{R}^n$ and $f : [a, b] \times \mathbb{R}^n \to \mathbb{R}^n$ (see Fig. 1.1). In words, the solution is a real-valued function and the differential equation means that its derivative is a real-valued function that depends on the independent variable (which is real) and the state variable.
- In the novel approach developed in this book, a solution to FIVP (1.2) is a fuzzy set in a space of functions, that is, $X(\cdot) \in \mathscr{F}(E([a, b]; \mathbb{R}^n))$, where $\mathscr{F}(\mathbb{X})$ denotes all the fuzzy sets of the universe $\mathbb{X}$. It means that each function has membership degree to the fuzzy set solution $X(\cdot)$. The differential equation is evaluated at each $t \in [a, b]$ and respective $X(t)$ with $F : [a, b] \times \mathscr{F}(\mathbb{R}^n) \to \mathscr{F}(\mathbb{R}^n)$. In other

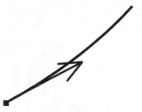

Fig. 1.1 Classical IVP: the solution is a function $x(\cdot) \in E([a, b]; \mathbb{R})$ and f is a function such that $f : [a, b] \times \mathbb{R} \to \mathbb{R}$

Fig. 1.2 FIVP using $\hat{D}$-derivative: the solution is a fuzzy bunch of functions $X(\cdot) \in \mathscr{F}(E([a, b]; \mathbb{R}))$ and F is a fuzzy-set-valued function such that $F : [a, b] \times \mathscr{F}(\mathbb{R}) \to \mathscr{F}(\mathbb{R})$

Fig. 1.3 FIVP via FDIs: the solution is a fuzzy bunch of functions $X(\cdot) \in \mathscr{F}(E([a, b]; \mathbb{R}))$ and F is a fuzzy-set-valued function such that $F : [a, b] \times \mathbb{R} \to \mathscr{F}(\mathbb{R})$

words, the derivative of the state variable X at t must equal F, which depends on the real independent variable t and the state variable at t, as illustrated in Fig. 1.2.

- A solution to FDIs is also of type $X(\cdot) \in \mathscr{F}(E([a, b]; \mathbb{R}^n))$, but the domain of the right-hand-side function is crisp, that is, $F : [a, b] \times \mathbb{R}^n \to \mathscr{F}(\mathbb{R}^n)$ (see Fig. 1.3). One does not solve an equation involving fuzzy sets but differential inclusions in the α-cuts.

Fig. 1.4 FIVP using H and GH-derivatives: the solution is a fuzzy-set-valued function $X(\cdot) \in E([a,b]; \mathscr{F}(\mathbb{R}))$ and F is such that $F : [a,b] \times \mathscr{F}(\mathbb{R}) \to \mathscr{F}(\mathbb{R})$

- The space of the solutions of the approach of Hukuhara derivative (or H-derivative) and the strongly generalized derivative (or GH-derivative) is another one: $X(\cdot) \in E([a,b]; \mathscr{F}(\mathbb{R}^n))$. In words, $X(\cdot)$ is a function that maps real values into fuzzy values. The right-hand-side function is of type $F : [a,b] \times \mathscr{F}(\mathbb{R}^n) \to \mathscr{F}(\mathbb{R}^n)$. The illustration of this approach is displayed in Fig. 1.4.
- Finally, the extension of the solution solves differential equations and extends the solution at each $t \in [a,b]$ such that $X(\cdot) \in E([a,b]; \mathscr{F}(\mathbb{R}^n))$.

The fuzziness in the solution enriches the theory of differential equations since the solutions are not composed of single points, but of sets of points associated with membership degrees. The following words are found in [12] where the author writes about multivalued functions (which is a particular case of fuzzy-set-valued functions):

> while to describe the behaviour of a point valued function is easy (a point can only displace itself), a set, besides displacing can be larger or smaller, can be convex or not. All these different facts are relevant to the problem of the existence of solutions and to their properties.

1.2.1 Historical Overview

The authors of [21] first used the term "fuzzy differential equations," with a completely different meaning from nowadays. They used and extended Zadeh's definition of the probability of a fuzzy event and solved differential equations involving the membership function of a given fuzzy set (which was not the unknown variable).

Later on, [33] defined the derivative for fuzzy functions based on the concept of Hukuhara derivative for set-valued functions. The first theorem of existence using this derivative was proposed by Kaleva [19], where the Lipschitz condition was used to assure existence and uniqueness of solution to a FIVP. With an equivalent derivative, in the same year [35] published similar result. Both explored the equivalence of the FIVP with a fuzzy integral equation using Aumann integral for fuzzy-set-valued functions proposed by Puri and Ralescu [34] (a generalization of the Aumann integral for set-valued functions). A version of Peano theorem of existence of solution was published by Kaleva [20]. Its author proved that the continuity of the function F in (1.2) and local compactness of the domain and the codomain of the state variable assured existence of solution to the FIVP.

Many other studies regarding solutions to FIVPs using the Hukuhara derivative were published. The authors in [38] proposed an existence and uniqueness theorem based on approximation by successive iterations. Local and global existence and uniqueness results for functional (or delay) differential equations were established in [27]. See also [7, 25, 26, 31].

The concept of Hukuhara derivative to solve FDE, in spite their use in many research articles, is considered to be defective since the differentiable functions have nondecreasing diameters. Therefore the solutions to differential equations cannot have decreasing diameter and, consequently, no periodic behavior can be modeled (nor can contractive behavior), except in the nonfuzzy case.

Based on the theory of differential inclusions for set-valued functions (interested readers can find good review and list of references in [12] and the main results in [2, 15]), [1, 3] proposed to solve FDIs. The idea is to solve differential inclusions considering the membership degrees for initial conditions, right-hand-side functions and solutions.

Reference [18] suggests to solve differential inclusions for each level of the right-hand side fuzzy function (a multivalued function). Using this interpretation, [14] proved an existence theorem for FDIs with fuzzy initial condition. It stated that if some hypotheses are met, the solutions of the differential inclusions produce a fuzzy bunch of functions, that is, a fuzzy set in a space of functions. Moreover, its attainable sets are fuzzy numbers.

The solution of an FDI may present decreasing diameter, overcoming the Hukuhara defect. This advantage and the richness of the fuzzy and the multivalued functions led many other authors to study this theory (see [4, 24, 25, 29, 39]).

This approach seems attractive, on the one hand, since it has no fuzzy derivatives. Hence it avoids the problem of solving equations with fuzzy sets, which looks much more complicated (minimization and maximization problems have to be frequently solved). On the other hand, solving differential inclusions is not an easy task as well. The extension of the solution of the IVP is intuitive and easier to solve (see [11, 29, 32]). It also preserves the main properties of the nonfuzzy case. Reference [30] proved that solutions that are stable for classical models are also stable for the fuzzy case. It does not use any fuzzy derivative, as in FDIs. In fact, under certain conditions, these two approaches are very similar and produce the same solutions for FIVPs.

The theory for the first generalizations of the Hukuhara derivative was proposed and developed in [8]. The strongly generalized and the weakly generalized differentiabilities are derivatives for fuzzy functions that differentiate all Hukuhara differentiable functions and others, including a class of functions with decreasing diameter. An existence theorem using the strongly generalized differentiability was also proven. This result assures two solutions to FIVPs, one for models of nonincreasing processes and the other one with nondecreasing diameter. A characterization result was obtained in [7] stating that the FDEs are equivalent to classical differential equation systems. That is, it is possible to solve many FIVPs by using only classical theory. Some interesting behaviors of solutions to FIVPs via generalized derivatives are novel in the field of differential equations. Phenomena such as "switch points," in which other solutions arise at determined points of the dynamics (even in very well-behaved dynamics), do not exist in classical theory.

Other more general derivatives—the generalized Hukuhara derivative and the most general so far, the fuzzy generalized derivative (see [36])—were suggested more recently. These generalized derivatives have been extensively studied (see [8, 10, 13, 36]). However, the development of the theory of FDEs using the generalizations of the Hukuhara differentiability has been limited to the strongly generalized version (see [8, 9, 22, 28, 37]).

Another derivative, namely the π-derivative, was extended from the case in which the functions are set-valued to the fuzzy-set-valued ones (see [13]). The π-derivative is based on the embedding of the family of nonempty compact sets of $\mathbb{R}^n$ in a real normed linear space.

The use of the extension principle to define the derivative and integral operators was suggested in [6]. These concepts were further investigated and an existence theorem for solutions to FIVPs was stated in [5]. The proof is based on the theorem of existence of solutions of FDI, revealing a connection between these two approaches. The mentioned theorem is part of this book, as well as other results connecting the extension of the derivative operator and via generalized derivatives. It will be clear that all approaches mentioned here have some similarities. Some of these results have already been published [5, 16, 17].

References

1. J.P. Aubin, Fuzzy differential inclusions. Probl. Control Inf. Theory **19**(1), 55–67 (1990)
2. J.P. Aubin, A. Cellina, *Differential Inclusions: Set-Valued Maps and a Viability Theory* (Springer, Berlin/Heidelberg, 1984)
3. V.A. Baidosov, Fuzzy differential inclusions. PMM USSR **54**(1), 8–13 (1990)
4. L.C. Barros, R.C. Bassanezi, R.Z.G. De Oliveira, Fuzzy differential inclusion: an application to epidemiology, in *Soft Methodology and Random Information Systems*, vol. 1, ed. by M. López-Díaz, M.A. Gil, P. Grzegorzewski, O. Hryniewicz, J. Lawry (Springer, Warsaw, 2004), pp. 631–637
5. L.C. Barros, L.T. Gomes, P.A. Tonelli, Fuzzy differential equations: an approach via fuzzification of the derivative operator. Fuzzy Sets Syst. **230**, 39–52 (2013)

6. L.C. Barros, P.A. Tonelli, A.P. Julião, Cálculo diferencial e integral para funções fuzzy via extensão dos operadores de derivação e integração (in Portuguese). Technical Report 6, 2010
7. B. Bede, Note on "numerical solutions of fuzzy differential equations by predictor-corrector method". Inf. Sci. **178**, 1917–1922 (2008)
8. B. Bede, S.G. Gal, Generalizations of the differentiability of fuzzy-number-valued functions with applications to fuzzy differential equations. Fuzzy Sets Syst. **151**, 581–599 (2005)
9. B. Bede, S.G. Gal, Solutions of fuzzy differential equations based on generalized differentiability. Commun. Math. Anal. **9**, 22–41 (2010)
10. B. Bede, L. Stefanini, Generalized differentiability of fuzzy-valued functions. Fuzzy Sets Syst. **230**, 119–141 (2013)
11. J.J. Buckley, T. Feuring, Almost periodic fuzzy-number-valued functions. Fuzzy Sets Syst. **110**, 43–54 (2000)
12. A. Cellina, A view on differential inclusions. Rend. Sem. Mat. Univ. Politech. Torino **63**, 197–209 (2005)
13. Y. Chalco-Cano, H. Román-Flores, M.D. Jiménez-Gamero, Generalized derivative and π-derivative for set-valued functions. Inf. Sci. **181**, 2177–2188 (2011)
14. P. Diamond, Time-dependent differential inclusions, cocycle attractors an fuzzy differential equations. IEEE Trans. Fuzzy Syst. **7**, 734–740 (1999)
15. A.F. Filippov, Differential equations with multi-valued discontinuous right-hand side. Dokl. Akad. Nauk SSSR **151**, 65–68 (1963)
16. L.T. Gomes, L.C. Barros, Fuzzy calculus via extension of the derivative and integral operators and fuzzy differential equations, in *2012 Annual Meeting of the North American Fuzzy Information Processing Society (NAFIPS)*, Berkeley (IEEE, 2012), pp. 1–5
17. L.T. Gomes, L.C. Barros, Fuzzy differential equations with arithmetic and derivative via Zadeh's extension. Mathware Soft Comput. Mag. **20**, 70–75 (2013)
18. E. Hüllermeier, An approach to modelling and simulation of uncertain dynamical systems. Int. J. Uncertainty Fuzziness Knowledge Based Syst. **5**(2), 117–137 (1997)
19. O. Kaleva, Fuzzy differential equations. Fuzzy Sets Syst. **24**, 301–317 (1987)
20. O. Kaleva, The Cauchy problem for fuzzy differential equations. Fuzzy Sets Syst. **35**, 389–396 (1990)
21. A. Kandel, W.J. Byatt, Fuzzy processes. Fuzzy Sets Syst. **4**, 117–152 (1980)
22. A. Khastan, J.J. Nieto, R. Rodríguez-López, Periodic boundary value problems for first-order linear differential equations with uncertainty under generalized differentiability. Inf. Sci. **222**, 544–558 (2013)
23. V. Křivan, Differential inclusions as a methodology tool in population biology, in *Proceedings of the 1995 European Simulation Multiconference* (Computer Simulation International, San Diego, 1995), pp. 544–547
24. V. Křivan, G. Colombo, A non-stochastic approach for modeling uncertainty in population dynamics. Bull. Math. Biol. **60**(4), 721–751 (1998)
25. V. Lakshmikantham, R. Mohapatra, *Theory of Fuzzy Differential Equations and Inclusions* (Taylor and Francis Publishers, London, 2003)
26. V. Lupulescu, Initial value problem for fuzzy differential equations under dissipative conditions. Inf. Sci. **178**, 4523–4533 (2009)
27. V. Lupulescu, On a class of fuzzy functional differential equations. Fuzzy Sets Syst. **160**, 1547–1562 (2009)
28. V. Lupulescu, J. Li, A. Zhao, J. Yan, The Cauchy problem of fuzzy differential equations under generalized differentiability. Fuzzy Sets Syst. **200**, 1–24 (2012)
29. M.T. Mizukoshi, L.C. Barros, Y. Chalco-Cano, H. Román-Flores, R.C. Bassanezi, Fuzzy differential equations and the extension principle. Inf. Sci. **177**, 3627–3635 (2007)
30. M.T. Mizukoshi, L.C. Barros, R.C. Bassanezi, Stability of fuzzy dynamic systems. Int. J. Uncertainty Fuzziness Knowledge Based Syst. **17**, 69–83 (2009)
31. J.J. Nieto, The Cauchy problem for continuous fuzzy differential equations. Fuzzy Sets Syst. **102**, 259–262 (1999)

32. M. Oberguggenberger, S. Pittschmann, Differential equations with fuzzy parameters. Math. Mod. Syst. **5**, 181–202 (1999)
33. M. Puri, D. Ralescu, Differentials of fuzzy functions. J. Math. Anal. Appl. **91**, 552–558 (1983)
34. M. Puri, D. Ralescu, Fuzzy random variables. J. Math. Anal. Appl. **114**, 409–422 (1986)
35. S. Seikkala, On the fuzzy initial value problem. Fuzzy Sets Syst. **24**, 309–330 (1987)
36. L. Stefanini, B. Bede, Generalized Hukuhara differentiability of interval-valued functions and interval differential equations. Nonlinear Anal. Theory Methods Appl. **71**, 1311–1328 (2009)
37. E.J. Villamizar-Roa, V. Angulo-Castillo, Y. Chalco-Cano, Existence of solutions to fuzzy differential equations with generalized Hukuhara derivative via contractive-like mapping principles. Fuzzy Sets Syst. **265**, 24–38 (2015)
38. C. Wu, S. Song, S. Lee, Approximate solutions, existence and uniqueness of the Cauchy problem of fuzzy differential equations. J. Math. Anal. Appl. **202**, 629–644 (1996)
39. Y. Zhu, L. Rao, Differential inclusions for fuzzy maps. Fuzzy Sets Syst. **112**, 257–261 (2000)
40. H.J. Zimmermann, Fuzzy set theory. Wiley Interdiscip. Rev. Comput. Stat. **2**(3), 317–332 (2010)

Chapter 2
Basic Concepts

This chapter introduces concepts such as fuzzy sets, extension principle, fuzzy numbers, α-cuts, fuzzy arithmetic and fuzzy metrics, and the notation we use in this text. They are fundamental for the reader who is not familiar with the theory of fuzzy sets in order to understand the following chapters. We also present famous and important results such as the Characterization Theorem, which characterizes α-cuts of fuzzy numbers as nonempty closed and bounded intervals. At the end of this chapter we present different kinds of fuzzy functions and means of comparing them, which will be needed when studying the solutions of differential equations under different approaches. For a deeper understanding, the reader can refer to [4, 8, 19, 30, 31] and the papers cited herein this book.

2.1 Fuzzy Subsets

Definition 2.1. A fuzzy subset A of a universe $\mathbb{U}$ is characterized by a function

$$\mu_A : \mathbb{U} \to [0, 1] \tag{2.1}$$

called membership function.

If

$$\mu_A : \mathbb{U} \to \{0, 1\} \tag{2.2}$$

the subset A is said to be crisp.

In the nonfuzzy case (2.2), μ_A is called the *characteristic function* (or *indicator function*) and it is often denoted by χ_A. If $\chi_A(x) = 0$, then x does not belong to A, whereas if $\chi_A(x) = 1$, then x belongs to A. The fuzzy subset is a generalization in which an element of $\mathbb{U}$ has partial membership to A characterized by a degree in the

© The Author(s) 2015
L.T. Gomes et al., *Fuzzy Differential Equations in Various Approaches*,
SpringerBriefs in Mathematics, DOI 10.1007/978-3-319-22575-3_2

interval $[0, 1]$. Hence the assignment $\mu_A(x) = 0$ means that x does not belong to A, while the closer $\mu_A(x)$ is to 1, the more x is considered in A.

Whenever a function, a set, or any other object is nonfuzzy, we refer to it just as "function," "set," and the like. If we believe it is necessary to stress that it is nonfuzzy, we use the words *crisp*, *classical*, or *nonfuzzy*.

Some important classical subsets related to fuzzy subsets are defined in what follows.

Definition 2.2. Given a fuzzy subset A of a topological space $\mathbb{U}$, its α-cuts (or α-levels) are the subsets

$$[A]_\alpha = \begin{cases} \{x \in \mathbb{U} : \mu_A(x) \geq \alpha\}, & \text{if } \alpha \in (0, 1] \\ \mathrm{cl}\,\{x \in \mathbb{U} : \mu_A(x) > 0\}, & \text{if } \alpha = 0 \end{cases} \tag{2.3}$$

where $\mathrm{cl}\,Z$ denotes the closure of the classical subset Z.

The support is

$$\mathrm{supp}\,A = \{x \in \mathbb{U} : \mu_A(x) > 0\}. \tag{2.4}$$

The core is

$$\mathrm{core}\,A = \{x \in \mathbb{U} : \mu_A(x) = 1\}. \tag{2.5}$$

Two fuzzy subsets A and B of $\mathbb{U}$ are said to be equal if their membership functions are the same for all element in $\mathbb{U}$ (i.e., $\mu_A(x) = \mu_B(x)$, $\forall x \in \mathbb{U}$). Or, equivalently, if all α-cuts coincide ($[A]_\alpha = [B]_\alpha$, for all $\alpha \in [0, 1]$).

We denote by

- $\mathcal{K}^n$ the family of all nonempty compact subsets of $\mathbb{R}^n$;
- $\mathcal{K}_{\mathscr{C}}^n$ the family of all nonempty compact and convex subsets of $\mathbb{R}^n$;
- $\mathscr{P}(\mathbb{U})$ the family of all subsets of $\mathbb{U}$;
- $\mathscr{F}(\mathbb{U})$ the family of all fuzzy sets of $\mathbb{U}$;
- $\mathscr{F}_{\mathscr{K}}(\mathbb{U})$ the family of fuzzy sets of $\mathbb{U}$ whose α-cuts are nonempty compact subsets of $\mathbb{U}$;
- $\mathscr{F}_{\mathscr{C}}(\mathbb{U})$ the family of fuzzy sets of $\mathbb{U}$ whose α-cuts are nonempty compact and convex subsets of $\mathbb{U}$.
- $E([a, b]; \mathbb{R}^n)$ a space of function from $[a, b]$ to $\mathbb{R}^n$, $a \leq b$, $a, b \in \mathbb{R}$. For instance, $C([a, b]; \mathbb{R}^n)$ is the space of continuous functions. Further examples appear in Appendix A.

If the core of a fuzzy subset is nonempty, the fuzzy subset is called *normal*. A fuzzy subset A of a vector space $\mathbb{U}$ is said to be *fuzzy convex* if $\mu_A(\lambda x + (1 - \lambda)y) \geq \min\{\mu_A(x), \mu_A(y)\}$ for every $x, y \in \mathbb{U}$, $\lambda \in [0, 1]$, that is, if its membership function is quasiconcave. If $\mathbb{U} = \mathbb{R}$, this condition assures that the α-cuts are intervals (convex subsets).

Fig. 2.1 A fuzzy subset in $\mathbb{R}$
that is a fuzzy number

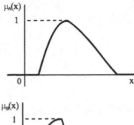

Fig. 2.2 A fuzzy subset in $\mathbb{R}$
that is not a fuzzy number

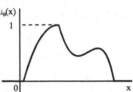

Definition 2.3. A fuzzy number is a fuzzy convex and normal fuzzy subset in $\mathbb{R}$ with upper semicontinuous membership function and compact support.

The family of the fuzzy numbers coincides with $\mathscr{F}_{\mathscr{C}}(\mathbb{R})$ (see Figs. 2.1 and 2.2 for examples of fuzzy sets in $\mathbb{R}$, one that is a fuzzy number and another that is not). Theorem 2.1 assures that all α-cuts of a fuzzy number are nonempty closed and bounded intervals with some properties whereas Theorem 2.2 is its converse, that is, if a family of nonempty closed intervals has some properties, they are the α-cuts of a unique fuzzy number. Hence, when dealing with fuzzy numbers it suffices to operate with their α-cuts; it is equivalent to operating with the fuzzy number itself.

Theorem 2.1 (Stacking Theorem, [28]). *A fuzzy number A satisfies the following conditions:*

(i) its α-cuts are nonempty closed intervals, for all $\alpha \in [0, 1]$;
(ii) if $0 \le \alpha_1 \le \alpha_2 \le 1$, then $[A]_{\alpha_2} \subseteq [A]_{\alpha_1}$;
(iii) for any nondecreasing sequence (α_n) in $[0, 1]$ converging to $\alpha \in (0, 1]$ we have

$$\bigcap_{n=1}^{\infty}[A]_{\alpha_n} = [A]_{\alpha};$$ (2.6)

and
(iv) for any nonincreasing sequence (α_n) in $[0, 1]$ converging to zero we have

$$cl\left(\bigcup_{n=1}^{\infty}[A]_{\alpha_n}\right) = [A]_0.$$ (2.7)

Theorem 2.2 (Characterization Theorem, [28]). *If $\{A_\alpha : \alpha \in [0, 1]\}$ is a family of subsets of $\mathbb{R}$ such that*

(i) A_α are nonempty closed intervals, for all $\alpha \in [0, 1]$;
(ii) if $0 \le \alpha_1 \le \alpha_2 \le 1$, then $A_{\alpha_2} \subseteq A_{\alpha_1}$;

(iii) for any nondecreasing sequence (α_n) in $[0, 1]$ converging to $\alpha \in (0, 1]$ we have

$$\bigcap_{n=1}^{\infty} A_{\alpha_n} = A_{\alpha}; \tag{2.8}$$

and
(iv) for any nonincreasing sequence (α_n) in $[0, 1]$ converging to zero we have

$$cl\left(\bigcup_{n=1}^{\infty} A_{\alpha_n}\right) = A_0, \tag{2.9}$$

then there exists a unique fuzzy number A such that $\{A_{\alpha} : \alpha \in [0, 1]\}$ are its α-cuts.

If A is a fuzzy number, we denote its α-cuts by $A_{\alpha} = [a_{\alpha}^-, a_{\alpha}^+]$ where a_{α}^- and a_{α}^+ are the lower and upper endpoints of the closed interval $[A]_{\alpha}$.

A particular kind of fuzzy number is the *triangular fuzzy number*. The notation for a triangular fuzzy number A with support $[a, c]$ and core $\{b\}$ is $(a; b; c)$ and its membership function is given by

$$\mu_A(x) = \begin{cases} \dfrac{x-a}{b-a}, & \text{if } x \in [a, b] \\ \dfrac{-x+c}{c-b}, & \text{if } x \in (b, c] \\ 0, & \text{if } x \notin [a, c] \end{cases} \tag{2.10}$$

where $a < b < c$ (see Fig. 2.3 for an example).

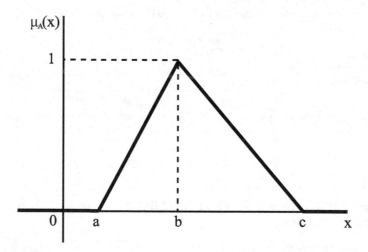

Fig. 2.3 Triangular fuzzy number

A particular set of fuzzy numbers is

$$\mathscr{F}_{\mathscr{C}}^0(\mathbb{R}) = \{A \in \mathscr{F}_{\mathscr{C}}(\mathbb{R}) : [A]_\alpha = [a_\alpha^-, a_\alpha^+], a_\cdot^-, a_\cdot^+ \in C([0, 1]; \mathbb{R})\}, \tag{2.11}$$

that is, the lower and upper endpoints of the level set functions of each fuzzy number are continuous in α. Some studies have been done using this particular class of fuzzy numbers (see, e.g., [4, 5]) and it will play an important role in connection with the Hukuhara and generalized differentiabilities and the $\hat{D}$-derivative (see Chaps. 3 and 4).

The definition of t-norm will be explored in the next section, to define a particular case of fuzzy arithmetic.

Definition 2.4. A t-norm is a function $T : [0, 1] \times [0, 1] \to [0, 1]$ that satisfies the following properties, for all $x, y, u, v \in [0, 1]$:

(i) Neutral element: $T(x, 1) = x$.
(ii) Commutativity: $T(x, y) = T(y, x)$.
(iii) Associativity: $T(x, T(y, z)) = T(T(x, y), z)$.
(iv) Monotonicity: if $x \le u$ and $y \le v$, then $T(x, y) \le T(u, v)$.

In order to deal with fuzzy subsets of general spaces, we present a generalization of Theorems 2.1 and 2.2 from the space $\mathbb{R}$ to more general topological spaces. Since we will deal not only with fuzzy subsets of $\mathbb{R}$, but with fuzzy subsets of spaces of functions as well, we state these results in what follows.

Theorem 2.3 ([7]). *Let $\mathbb{X}$ be a topological space. A fuzzy subset $A \in \mathscr{F}_\mathscr{X}(\mathbb{X})$ satisfies the following conditions:*

(i) if $0 \le \alpha_1 \le \alpha_2 \le 1$, then $[A]_{\alpha_2} \subseteq [A]_{\alpha_1}$;
(ii) for any nondecreasing sequence (α_n) in $[0, 1]$ converging to $\alpha \in (0, 1]$ we have

$$\cap_{n=1}^\infty [A]_{\alpha_n} = [A]_\alpha; \tag{2.12}$$

and
(iii) for any nonincreasing sequence (α_n) in $[0, 1]$ converging to zero we have

$$cl\left(\cup_{n=1}^\infty [A]_{\alpha_n}\right) = [A]_0. \tag{2.13}$$

Theorem 2.4 ([7]). *Let $\mathbb{X}$ be a topological space. If $\{A_\alpha : \alpha \in [0, 1]\}$ is a family of subsets of $\mathbb{X}$ such that*

(i) A_α are nonempty compact subsets, for all $\alpha \in [0, 1]$;
(ii) if $0 \le \alpha_1 \le \alpha_2 \le 1$, then $A_{\alpha_2} \subseteq A_{\alpha_1}$;
(iii) for any nondecreasing sequence (α_n) in $[0, 1]$ converging to $\alpha \in (0, 1]$ we have

$$\cap_{n=1}^\infty A_{\alpha_n} = A_\alpha; \tag{2.14}$$

and

(iv) for any nonincreasing sequence (α_n) in $[0, 1]$ converging to zero we have

$$cl \left(\cup_{n=1}^{\infty} A_{\alpha_n}\right) = A_0, \tag{2.15}$$

then there exists a unique fuzzy subset $A \in \mathscr{F}_{\mathscr{K}}(\mathbb{X})$ such that $\{A_\alpha : \alpha \in [0, 1]\}$ are its α-cuts.

We finish this section by presenting the linear structure in $\mathscr{F}_{\mathscr{K}}(\mathbb{R}^n)$ used in the literature.

Consider the fuzzy subsets $A, B \in \mathscr{F}_{\mathscr{K}}(\mathbb{R}^n)$ and $\lambda \in \mathbb{R}$,

$$\mu_{A+B}(z) = \sup_{x+y=z} \min\{\mu_A(x), \mu_B(y)\} \tag{2.16}$$

and

$$\mu_{\lambda A}(z) = \begin{cases} \mu_A(z/\lambda), & \text{if } \lambda \neq 0 \\ \chi_0(z), & \text{if } \lambda = 0 \end{cases} \tag{2.17}$$

where we write $\chi_0(z)$ instead of $\chi_{\{0\}}(z)$ for simplification.

From the theory presented in the next section, one proves that

$$[A + B]_\alpha = [A]_\alpha + [B]_\alpha \quad \text{and} \quad [\lambda A]_\alpha = \lambda[A]_\alpha, \tag{2.18}$$

where

$$[A]_\alpha + [B]_\alpha = \{a + b : a \in [A]_\alpha, b \in [B]_\alpha\} \tag{2.19}$$

is the Minkowski sum of $[A]_\alpha$ and $[B]_\alpha$ and

$$\lambda[A]_\alpha = \{\lambda a : a \in [A]_\alpha\}. \tag{2.20}$$

2.2   Extension Principle

Zadeh's extension principle (see [29, 36]), to which we refer as *extension principle* in the rest of the text, is the fuzzy version of the *united extension* (according to [22]), that extends functions whose inputs and outputs are points to functions whose inputs and outputs are sets. Given a classical function f and a classical subset A as input, the united extension $\hat{f}(A)$ is defined as the union of the images of all elements of A (see Fig. 2.4). In the fuzzy case, it is intuitive that if a membership degree $\mu_A(x)$ is assigned to one element x in a subset A, the image $y = f(x)$ of this element by means of an injective function has the same membership degree $\mu_{\hat{f}(A)}(y) = \mu_A(x)$ (see Fig. 2.5). If an element has more than one preimage, its membership degree is given by the supremum of the membership degrees of all possible preimages. This process

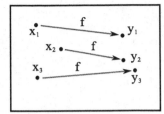
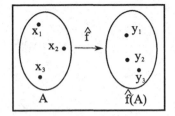

Fig. 2.4 United extension of a function f on a classical subset A: f evaluated at each element of A defines the united extension of f at A

Fig. 2.5 Extension of a function f on a fuzzy subset A: f evaluated at each element of A, together with its membership degree to A, defines the extension $\hat{f}(A)$

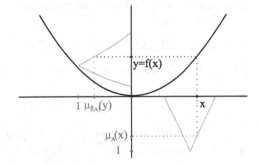

of extending a function is known as *extension principle* and it is defined in the fuzzy context in Definitions 2.5 and 2.6, where the former is a particular case of the latter.

Definition 2.5 (Extension Principle [29, 36]). Let $\mathbb{U}$ and $\mathbb{V}$ be two universes and $f : \mathbb{U} \to \mathbb{V}$ a classical function. For each $A \in \mathscr{F}(\mathbb{U})$ we define the extension of f as $\hat{f}(A) \in \mathscr{F}(\mathbb{V})$ such that

$$\mu_{\hat{f}(A)}(y) = \begin{cases} \sup_{s \in f^{-1}(y)} \mu_A(s), & \text{if } f^{-1}(y) \neq \emptyset \\ 0, & \text{if } f^{-1}(y) = \emptyset \end{cases}, \tag{2.21}$$

for all $y \in \mathbb{V}$, where $f^{-1}(y) = \{x \in \mathbb{U} : f(x) = y\}$.

Example 2.1. Let

$$f(x) = ax + b \tag{2.22}$$

with $a, b \in \mathbb{R}$, $a \neq 0$. Since $f^{-1}(y) = a^{-1}(y - b)$, the extension of f is the fuzzy function $\hat{f}$ such that, given $X \in \mathscr{F}(\mathbb{R})$,

$$\mu_{\hat{f}(X)}(y) = \sup_{x = a^{-1}(y-b)} \mu_X(x) = \mu_X(a^{-1}(y - b)) \tag{2.23}$$

for all $y \in \mathbb{R}$. Or

$$\mu_{\hat{f}(X)}(ax + b) = \mu_X(x) \tag{2.24}$$

that is,

$$\hat{f}(X) = aX + b. \tag{2.25}$$

The next theorem allows us to determine the extension of continuous and/or surjective functions more easily. It states that it suffices to calculate the image of the function on each element of the α-cut of the argument, in order to obtain the α-cut of the image.

Theorem 2.5 ([1, 29]). *Let $f : \mathbb{R}^n \to \mathbb{R}^m$ be a function.*

(a) If f is surjective, then a necessary and sufficient condition for

$$[\hat{f}(A)]_\alpha = f([A]_\alpha) \tag{2.26}$$

to hold is that $\sup\{\mu_A(x) : x \in f^{-1}(y)\}$ be attained for each $y \in \mathbb{R}^m$.
(b) If f is continuous, then $\hat{f} : \mathscr{F}_{\mathscr{K}}(\mathbb{R}^n) \to \mathscr{F}_{\mathscr{K}}(\mathbb{R}^n)$ is well defined and

$$[\hat{f}(A)]_\alpha = f([A]_\alpha) \tag{2.27}$$

for all $\alpha \in [0, 1]$.

A generalization of item (b) and its proof can be found in [7].

Theorem 2.6 ([7]). *Let $\mathbb{U}$ and $\mathbb{V}$ be two Hausdorff spaces and $f : \mathbb{U} \to \mathbb{V}$ be a function. If f is continuous, then $\hat{f} : \mathscr{F}_{\mathscr{K}}(\mathbb{U}) \to \mathscr{F}_{\mathscr{K}}(\mathbb{V})$ is well defined and*

$$[\hat{f}(A)]_\alpha = f([A]_\alpha) \tag{2.28}$$

for all $\alpha \in [0, 1]$.

Proof (Adapted from [7]). From Definition 2.21, $\hat{f}(A)$ is a fuzzy subset in $\mathbb{V}$. To prove that $\hat{f} : \mathscr{F}_{\mathscr{K}}(\mathbb{U}) \to \mathscr{F}_{\mathscr{K}}(\mathbb{V})$, it is needed to prove that the α-cuts $[\hat{f}(A)]_\alpha$ are nonempty compact subsets of $\mathbb{V}$. Since f is continuous, it assigns compact subsets to compact subsets, hence it suffices to prove Equation $[\hat{f}(A)]_\alpha = f([A]_\alpha)$.

Equation (2.28) will be proved. First we show that

(i) $f([A]_\alpha) \subseteq [\hat{f}(A)]_\alpha$. Consider $A \in \mathscr{F}_{\mathscr{K}}(\mathbb{U})$ and $y \in f([A]_\alpha)$. Then there exists $x \in [A]_\alpha$ such that $y = f(x)$. From Definition 2.21 (extension principle), $\mu_{\hat{f}(A)}(y) = \sup_{x \in f^{-1}(y)} \mu_A(x) \geq \alpha$. Hence $y \in [\hat{f}(A)]_\alpha$ and the conclusion is $f([A]_\alpha) \subseteq [\hat{f}(A)]_\alpha$. Now we show that

(ii) $[\hat{f}(A)]_\alpha \subseteq f([A]_\alpha)$. Let us remark first that, since $\mathbb{U}$ and $\mathbb{V}$ are Hausdorff spaces, a single point $y \in \mathbb{V}$ is closed. Moreover, the continuity of f implies that $f^{-1}(y)$ is

closed. Since $[A]_0$ is compact, $f^{-1}(y) \cap [A]_0$ is also compact. For $\alpha > 0$, consider $y \in [\hat{f}(A)]_\alpha$. Then $\mu_{\hat{f}(A)}(y) = \sup_{x \in f^{-1}(y)} \mu_A(x) \geq \alpha > 0$ and, therefore, there exist $x \in f^{-1}(y)$ such that $f^{-1}(y) \cap [A]_0 \neq \varnothing$.

Also, since $\mu_A(x)$ is upper semicontinuous and $f^{-1}(y) \cap [A]_0$ is compact, the supremum is attained, that is, there exists $x \in f^{-1}(y) \cap [A]_0$ with $\mu_{\hat{f}(A)}(y) = \mu_A(x) \geq \alpha$. Hence $y = f(x)$ for some $x \in [A]_\alpha$, that is, $y \in f([A]_\alpha)$.

For $\alpha = 0$, the results obtained yield

$$\bigcup_{\alpha \in (0,1]} [\hat{f}(A)]_\alpha = \bigcup_{\alpha \in (0,1]} f([A]_\alpha) \subseteq f([A]_0). \tag{2.29}$$

Since $f([A]_0)$ is closed,

$$[\hat{f}(A)]_0 = cl\left(\bigcup_{\alpha \in (0,1]} [\hat{f}(A)]_\alpha\right) = cl\left(\bigcup_{\alpha \in (0,1]} f([A]_\alpha)\right) \subseteq f([A]_0). \tag{2.30}$$

Hence $[\hat{f}(A)]_\alpha \subseteq f([A]_\alpha)$ for all $\alpha \in [0, 1]$ and (2.28) follows from *(i)* and *(ii)*.

A fuzzy-set-valued function whose domain is not fuzzy is extended using Definition 2.6. It is a more general case than Definition 2.5 and it has a wider application.

Definition 2.6 (Extension Principle [29, 36]). Let $\mathbb{U}$ and $\mathbb{V}$ be two topological spaces and $F : \mathbb{U} \to \mathscr{F}(\mathbb{V})$ a function. For each $A \in \mathscr{F}(\mathbb{U})$ we define the extension of F as $\hat{F}(A) \in \mathscr{F}(\mathbb{V})$ where its (unique) membership function is given by

$$\mu_{\hat{F}(A)}(y) = \sup_{x \in \mathbb{U}} \{\mu_{F(x)}(y) \wedge \mu_A(x)\}. \tag{2.31}$$

for all $y \in \mathbb{V}$.

2.3 Fuzzy Arithmetics for Fuzzy Numbers

The α-cuts of fuzzy numbers are closed intervals so it is inevitable the influence of the concepts of the interval arithmetic on the arithmetic of fuzzy numbers. The first fuzzy arithmetic approach presented in this study is equivalent to the interval arithmetic with α-cuts of fuzzy numbers.

2.3.1   Standard Interval Arithmetic and Extension Principle

The standard interval arithmetic (SIA) [27] can be regarded as the united extension of the operators addition $(+)$, subtraction $(-)$, multiplication $(\cdot)$, and division $(\div)$ between real numbers. For instance, the addition of two intervals $A = [a^-, a^+]$ and $B = [b^-, b^+]$ is defined by applying the operation "addition" on every single pair $(a, b) \in A \times B$, that is,

$$A + B = \{a + b : a \in A, b \in B\} \tag{2.32}$$

The other three operations are defined likewise, i.e,

$$
\begin{aligned}
A - B &= \{a - b : a \in A, b \in B\} \\
A \cdot B &= \{a \cdot b : a \in A, b \in B\} \\
A \div B &= \{a \div b : a \in A, b \in B, 0 \notin B\}.
\end{aligned}
\tag{2.33}
$$

It is obvious from the definition of SIA that the arithmetic for real numbers is a particular case.

The fuzzy arithmetic based on SIA is the application of SIA on the α-cuts of two fuzzy numbers. It is equivalent to the proposal in [26] of the extension of the arithmetic operators, defined for real numbers. Given an arithmetic operator $\odot \in \{+, -, \cdot, \div\}$ and two fuzzy numbers A and B, the extension principle gives

$$\mu_{A \odot B}(c) = \sup_{a \odot b = c} \min\{\mu_A(a), \mu_B(b)\}. \tag{2.34}$$

Since the arithmetic operators are continuous functions, they are equivalent to operating on the elements of the α-cuts.

Consider two fuzzy numbers A and B with α-cuts $[A]_\alpha = [a_\alpha^-, a_\alpha^+]$ and $[B]_\alpha = [b_\alpha^-, b_\alpha^+]$. Using the extension principle (Definition 2.5), levelwise the sum is equivalent to

$$[A + B]_\alpha = [a_\alpha^- + b_\alpha^-, a_\alpha^+ + b_\alpha^+], \tag{2.35}$$

the subtraction is

$$[A - B]_\alpha = [a_\alpha^- - b_\alpha^+, a_\alpha^+ - b_\alpha^-], \tag{2.36}$$

the product is

$$[A \cdot B]_\alpha = \left[\min_{s, r \in \{-, +\}} a_\alpha^s \cdot b_\alpha^r, \max_{s, r \in \{-, +\}} a_\alpha^s \cdot b_\alpha^r \right], \tag{2.37}$$

and the division is

$$[A \div B]_\alpha = \left[\min_{s,r\in\{-,+\}} \left\{ \frac{a_\alpha^s}{b_\alpha^r} \right\}, \ \max_{s,r\in\{-,+\}} \left\{ \frac{a_\alpha^s}{b_\alpha^r} \right\} \right], \ 0 \notin \operatorname{supp} B. \qquad (2.38)$$

As mentioned above, this is the same as interval arithmetic on α-cuts.

Note that the difference between two identical nonzero width intervals is never the number zero. The difference in the limit

$$\lim_{h\to 0^+} \frac{F(x+h) - F(x)}{h} \qquad (2.39)$$

of a constant non-crisp function F is a constant non-crisp fuzzy number. The division by a variable tending towards zero is not defined. Therefore, to define the derivative of a fuzzy-number-valued function with the above arithmetic leads to a serious shortcoming.

This problem in defining the derivative happens due to the fact that this arithmetic takes into account every possible result. The sum is the same as Minkowski sum, in which all elements of a subset are added to all elements of the other subset, generating the largest possible subset as a result. The same happens to the subtraction. There are some approaches to overcome this, considering some kind of dependency between the variables.

2.3.2 Interactive Arithmetic

The addition of interactive fuzzy numbers using the generalization of the extension principle via t-norms (see [10]) provides a means of controlling the growth of uncertainty in calculations, differently from the arithmetic via traditional extension principle [13]. To define interactivity, the concept of joint membership function (analogous to joint possibility distribution, from the possibility theory) is needed.

Definition 2.7. If A_1 and A_2 are two fuzzy numbers, C is said to be their joint membership function if

$$\mu_{A_i}(a_i) = \max_{a_j \in \mathbb{R}, j\neq i} \mu_C(a_1, a_2). \qquad (2.40)$$

Two fuzzy numbers A_1 and A_2 are said to be noninteractive if their joint membership function satisfies

$$\mu_C(a_1, a_2) = \min\{\mu_{A_1}(a_1), \mu_{A_2}(a_2)\}. \qquad (2.41)$$

In words, the joint membership function is given by the t-norm of minimum. Otherwise, they are said to be interactive.

The generalization in [10] admits that any t-norm T can replace the min operator in (2.34). Reference [6] generalizes it even more:

$$\mu_{\hat{f}(A_1, A_2)}(c) = \begin{cases} \sup_{(a_1, a_2) \in f^{-1}(c)} \mu_C(a_1, a_2), & \text{if } f^{-1}(c) \neq \varnothing \\ 0, & \text{if } f^{-1}(c) = \varnothing \end{cases}. \tag{2.42}$$

Addition, subtraction, multiplication, and division are obtained in [12] extending the respective classical operators via (2.42) where the joint membership are t-norms. A particular case of joint membership is used in [6] to define addition and subtraction of interactive fuzzy numbers. It is based on completely correlated fuzzy numbers, i.e., given two fuzzy numbers A_1 and A_2, their joint membership is

$$\mu_C(a_1, a_2) = \mu_{A_1}(a_1) \cdot \chi_{\{qa_1 + r = a_2\}}(a_1, a_2) = \mu_{A_2}(a_2) \cdot \chi_{\{qa_1 + r = a_2\}}(a_1, a_2), \tag{2.43}$$

where $\chi_{\{qa_1 + r = a_2\}}$ is the characteristic function of the line

$$\{(a_1, a_2) \in \mathbb{R}^2 | qa_1 + r = a_2\}. \tag{2.44}$$

2.3.3   Constraint Interval Arithmetic

The constraint interval arithmetic (CIA), which deals with dependencies, redefines intervals as single-valued functions [21]. That is, an interval $[a^-, a^+]$ is given by the function $A^I(a^-, a^+, \lambda_A) = \{a : a = (1 - \lambda_A)a^- + \lambda_A a^+, 0 \leq \lambda_A \leq 1\}$.

Addition, multiplication, subtraction, and division between two intervals $A = [a^-, a^+]$ and $B = [b^-, b^+]$ are given by the formula

$$A \circ B = \{z : [(1 - \lambda_A)a^- + \lambda_A a^+] \circ [(1 - \lambda_B)b^- + \lambda_A a^+], 0 \leq \lambda_A \leq 1, 0 \leq \lambda_B \leq 1\} \tag{2.45}$$

where $\circ$ stands for any of the four arithmetic operations. In the case in which the two variables are the same,

$$A \circ A = \{z : [(1 - \lambda_A)a^- + \lambda_A a^+] \circ [(1 - \lambda_A)a^- + \lambda_A a^+], 0 \leq \lambda_A \leq 1\} \tag{2.46}$$

where, unlike SIA, $A - A = [0, 0]$ and $A \div A = [1, 1]$ (see [23]).

This idea is also defined to the fuzzy case [24], based on the affirmation that arithmetic of fuzzy numbers is arithmetic of intervals in each α-cut. They also have demonstrated that this arithmetic for fuzzy numbers is the same as gradual number arithmetic (see [11]).

This approach can be interpreted as a particular case of the arithmetic presented in Sect. 2.3.2, where A is completely correlated to A with $q = 1$ and A and B are noninteractive.

2.3.4 Hukuhara and Generalized Differences

The Hukuhara difference for intervals was defined in order to overcome the fact that $A - A \neq \{0\}$ for any interval A (see [17]) and it was used to define the Hukuhara difference for elements of $\mathscr{F}_{\mathscr{C}}(\mathbb{R})$ [32].

Definition 2.8 ([32]). Given two fuzzy numbers, $A, B \in \mathscr{F}_{\mathscr{C}}(\mathbb{R})$ the Hukuhara difference (H-difference for short) $A \ominus_H B = C$ is the fuzzy number C such that $A = B + C$, if it exists.

Levelwise,

$$[A \ominus_H B]_\alpha = [a_\alpha^- - b_\alpha^-, a_\alpha^+ - b_\alpha^+] \tag{2.47}$$

for all $\alpha \in [0, 1]$.

The Hukuhara difference has the property $A \ominus_H A = \{0\}$. However, this difference is not defined for pairs of fuzzy numbers such that the support of a fuzzy number has bigger diameter than the one that is subtracted. Two other definitions for difference of fuzzy numbers generalize the Hukuhara difference and are stated next [34, 35].

Definition 2.9 ([34, 35]). Given two fuzzy numbers $A, B \in \mathscr{F}_{\mathscr{C}}(\mathbb{R})$, the generalized Hukuhara difference (gH-difference for short) $A \ominus_{gH} B = C$ is the fuzzy number C, if it exists, such that

$$\begin{cases} (i) & A = B + C \text{ or} \\ (ii) & B = A - C. \end{cases} \tag{2.48}$$

Definition 2.10 ([5, 34]). Given two fuzzy numbers $A, B \in \mathscr{F}_{\mathscr{C}}(\mathbb{R})$, the generalized difference (g-difference for short) $A \ominus_g B = C$ is the fuzzy number C, if it exists, with α-cuts

$$[A \ominus_g B]_\alpha = \mathrm{cl} \bigcup_{\beta \geq \alpha} ([A]_\beta \ominus_{gH} [B]_\beta), \forall \alpha \in [0, 1], \tag{2.49}$$

where the gH-difference $\ominus_{gH}$ is with interval operands $[A]_\beta$ and $[B]_\beta$.

Example 2.2. The fuzzy numbers A and B with membership functions defined by

$$\mu_A(x) = \begin{cases} x + 1, & \text{if } x \in [-1, 0], \\ -x + 1, & \text{if } x \in (0, 1], \\ 0, & \text{otherwise} \end{cases} \quad \mu_B(x) = \begin{cases} 1, & \text{if } x \in [-1, 1], \\ 0, & \text{otherwise} \end{cases} \tag{2.50}$$

have, as gH-difference levelwise,

$$[A \ominus_{gH} B]_\alpha = [-\alpha, \alpha], \tag{2.51}$$

for all $\alpha \in [0, 1]$. This is not a fuzzy number. But for the g-difference we have

$$[A \ominus_g B]_\alpha = \text{cl} \bigcup_{\beta \geq \alpha} [-\beta, \beta]$$

$$= [-1, 1]$$

(2.52)

for all $\alpha \in [0, 1]$ and this is a fuzzy number.

Example 2.2 illustrates that the Definition 2.10 is more general than Definition 2.9, that is, it is defined for more pairs of fuzzy numbers. This means that whenever the gH-difference exists the g-difference exists and it is the same. In terms of α-cuts we have

$$[A \ominus_{gH} B]_\alpha = [\min\{a_\alpha^- - b_\alpha^-, a_\alpha^+ - b_\alpha^+\}, \max\{a_\alpha^- - b_\alpha^-, a_\alpha^+ - b_\alpha^+\}]$$

(2.53)

and

$$[A \ominus_g B]_\alpha = \left[\inf_{\beta \geq \alpha} \min\{a_\beta^- - b_\beta^-, a_\beta^+ - b_\beta^+\}, \sup_{\beta \geq \alpha} \max\{a_\beta^- - b_\beta^-, a_\beta^+ - b_\beta^+\} \right].$$

(2.54)

for all $\alpha \in [0, 1]$.

The g-difference is not defined for every pair of fuzzy numbers, though [15]. But among the differences that generalize the H-difference, it is the most general one proposed so far. This possibility of non-existence of the g-difference is illustrated in the next example.

Example 2.3. Consider the fuzzy numbers A and B with membership functions defined by

$$\mu_A(x) = \begin{cases} 1, & \text{if } x \in [2, 3], \\ 0.5, & \text{if } x \in [0, 2) \cup (3, 5], \\ 0, & \text{otherwise} \end{cases}$$

(2.55)

and

$$\mu_B(x) = \begin{cases} 1, & \text{if } x \in [2, 3], \\ 0.5, & \text{if } x \in [-1, 2) \cup (3, 4], \\ 0, & \text{otherwise.} \end{cases}$$

(2.56)

The gH-difference levelwise is

$$[A \ominus_{gH} B]_\alpha = \begin{cases} \{0\}, & \text{if } 0.5 < \alpha \leq 1, \\ \{1\}, & \text{if } 0 \leq \alpha \leq 0.5. \end{cases}$$

(2.57)

Hence we have the g-difference levelwise

$$[A \ominus_g B]_\alpha = \begin{cases} \{0\}, & \text{if } 0.5 < \alpha \le 1, \\ \{0\} \cup \{1\}, & \text{if } 0 \le \alpha \le 0.5, \end{cases} \tag{2.58}$$

which is not a fuzzy number.

Summary of Hukuhara and generalized differences:

- **The gH- and g-differences generalize the H-difference.** If the H-difference between two fuzzy numbers exists, the gH- and g-differences exist and they all have the same value.
- **The g-difference generalizes the gH-difference.** If the gH-difference between two fuzzy numbers exists, the g-difference exists and they have the same value.
- **The Hukuhara and the generalized differences do not always exist.** The H-, gH-, and g-differences between two fuzzy numbers do not always exist.

2.4 Fuzzy Metric Spaces

This section reviews some important definitions and results regarding fuzzy metric spaces. They can be found, together with proofs, in several references, e.g. [4, 8, 32].

The most used metric for fuzzy numbers is the Pompeiu–Hausdorff, based on Pompeiu–Hausdorff distance for compact convex subsets of a metric space $\mathbb{U}$. It is in turn based on the concept of Hausdorff separation.

Definition 2.11. Let A and B be two nonempty compact subsets of a metric space $\mathbb{U}$. The pseudometric

$$\rho(A, B) = \sup_{a \in A} d(a, B), \tag{2.59}$$

where

$$d(a, B) = \inf_{b \in B} ||a - b|| \tag{2.60}$$

is called Hausdorff separation.

Definition 2.12. Let A and B be two nonempty compact subsets of a metric space $\mathbb{U}$. The Pompeiu–Hausdorff metric d_H is given by

$$d_H(A, B) = \max\{\rho(A, B), \rho(B, A)\}. \tag{2.61}$$

For the space $\mathscr{F}_{\mathscr{K}}(\mathbb{U})$ (recall that the space of fuzzy numbers $\mathscr{F}_{\mathscr{C}}(\mathbb{R})$ is a particular case where $\mathbb{U} = \mathbb{R}$), the Pompeiu–Hausdorff metric is defined as follows.

Definition 2.13. Let A and B be elements of $\mathscr{F}_{\mathscr{K}}(\mathbb{U})$, where $\mathbb{U}$ is a metric space. The Pompeiu–Hausdorff metric d_∞ is defined as

$$d_\infty(A, B) = \sup_{\alpha \in [0,1]} d_H([A]_\alpha, [B]_\alpha). \qquad (2.62)$$

In the case of fuzzy numbers, that is, $A, B \in \mathscr{F}_{\mathscr{C}}(\mathbb{R})$, $d_\infty(A, B)$ is rewritten as

$$d_\infty(A, B) = \sup_{\alpha \in [0,1]} \max\{|a_\alpha^- - b_\alpha^-|, |a_\alpha^+ - b_\alpha^+|\}. \qquad (2.63)$$

Another known metrics are the endographic and the L^p-type distances.

Definition 2.14. Let A and B be elements of $\mathscr{F}_{\mathscr{K}}(\mathbb{U})$, where $\mathbb{U}$ is a metric space. The endographic metric d_E is defined as

$$d_E(A, B) = d_H(\text{send}(A), \text{send}(B)), \qquad (2.64)$$

where

$$\text{send}(A) = ([A]_0 \times [0, 1]) \cap \text{end}(A) \qquad (2.65)$$

with

$$\text{end}(A) = \{(x, \alpha) \in \mathbb{R}^n \times [0, 1] : \mu_A(x) \geq \alpha\}. \qquad (2.66)$$

Definition 2.15. Let A and B be elements of $\mathscr{F}_{\mathscr{K}}(\mathbb{U})$, where $\mathbb{U}$ is a metric space. The d_p distance is defined as

$$d_p(A, B) = \left(\int_0^1 d_H([A]_\alpha, [B]_\alpha)^p d\alpha \right)^{1/p}. \qquad (2.67)$$

We denote by $\overline{B}(X, q)$ the closed ball

$$\overline{B}(X, q) = \{A \in \mathscr{F}_{\mathscr{C}}(\mathbb{U}) : d_\infty(X, A) \leq q\}. \qquad (2.68)$$

The following theorem is a well-known result.

Theorem 2.7 ([33]). *The space of fuzzy numbers endowed with the d_∞ metric, denoted $(\mathscr{F}_{\mathscr{C}}(\mathbb{R}), d_\infty)$, is a complete metric space.*

Note that $(\mathscr{F}_{\mathscr{C}}(\mathbb{R}), d_\infty)$ is not separable. There exist other metrics that make the space of fuzzy numbers separable, but not complete (e.g., d^p with $1 \leq p < \infty$, see [4] and d_E [1, 20]).

Another important result is the Embedding Theorem. It connects the space of fuzzy numbers to a subset of pairs of functions, that define the endpoints of the α-cuts of fuzzy numbers. In other words, it allows us to use a well-known theory and tools for real functions, instead of operating with fuzzy numbers, which is more complicated. A general version of the theorem is for the space $\mathscr{F}_{\mathscr{C}}(\mathbb{R}^n)$ and is as follows.

Theorem 2.8 (Embedding Theorem, [18, 25, 32]). *There exists a real Banach space $\mathbb{X}$ such that the metric space $(\mathscr{F}_{\mathscr{C}}(\mathbb{R}^n), d_\infty)$ can be embedded isometrically into $\mathbb{X}$.*

Another application of the d_∞ metric is a result analogous to Theorem 2.5 (b), stated in [16]. It regards continuity of fuzzy-number-valued functions and extension principle. The concept of fuzzy function will be further explored in the next section.

Theorem 2.9 ([16]). *Let $F : \mathbb{R} \to \mathscr{F}_{\mathscr{C}}(\mathbb{R})$ be a d_∞-continuous function. Then the extension $\hat{F} : \mathscr{F}_{\mathscr{C}}(\mathbb{R}) \to \mathscr{F}_{\mathscr{C}}(\mathbb{R})$ is well defined, is d_∞-continuous, and*

$$[\hat{F}(A)]_\alpha = \bigcup_{a \in [A]_\alpha} [F(a)]_\alpha \tag{2.69}$$

for all $\alpha \in [0, 1]$.

Example 2.4. Let

$$F(x) = \Lambda x \tag{2.70}$$

with $\Lambda \in \mathscr{F}_{\mathscr{C}}(\mathbb{R})$. Then F is a d_∞-continuous function and $F : \mathbb{R} \to \mathscr{F}_{\mathscr{C}}(\mathbb{R})$. Applying Theorem 2.9,

$$[\hat{F}(X)]_\alpha = \bigcup_{x \in [X]_\alpha} [\Lambda x]_\alpha = \bigcup_{x \in [X]_\alpha} x[\Lambda]_\alpha = [X]_\alpha [\Lambda]_\alpha = [\Lambda X]_\alpha \tag{2.71}$$

where multiplication between intervals and multiplication between fuzzy numbers is the one defined in Sect. 2.3.1 (SIA).

As a result,

$$\hat{F}(X) = \Lambda X. \tag{2.72}$$

Since the mentioned results are important in this study, whenever we treat limits of sequences of fuzzy subsets or continuity of fuzzy-set-valued functions, it will assumed to be with respect to the d_∞ metric, unless another distance is specified.

2.5 Fuzzy Functions

Both the fuzzy bunches of functions and the fuzzy-set-valued functions are called fuzzy functions in [9]. The same is done in this text. In the literature in general, operations such as differentiation and integration are defined only for mappings from a classical space to a fuzzy space, called *fuzzy-set-valued function*. A mapping from a fuzzy space to another fuzzy space is more general and it appears in FIVPs as the function of the right-hand-side of the FDE. A *fuzzy-number-valued function* is a more restricted case: it takes classical points to fuzzy numbers. Fuzzy-set-valued functions are generalizations of *set-valued functions*. A set-valued function on I is a mapping $G : I \rightarrow \mathscr{P}(\mathbb{X})$ such that $G(t) \neq \varnothing$ for all $t \in I$, where I is usually an interval. That is, it takes points of I to the powerset of $\mathbb{X}$.

A *fuzzy bunch of functions* (or fuzzy bunch, for short) is a fuzzy subset of a space of functions. To be precise, it is not a function, but it is used to define solutions to fuzzy initial value problems. Also, to each fuzzy bunch of functions there corresponds a fuzzy-set-valued function, via *attainable fuzzy sets*. For each fuzzy bunch $F \in \mathscr{F}(E(I; \mathbb{R}^n))$, where $E(I; \mathbb{R}^n)$ is a space of functions from $I \subseteq \mathbb{R}$ to $\mathbb{R}^n$, the *attainable fuzzy sets* at t, $F(t)$, are the fuzzy sets of $\mathbb{R}^n$ such that

$$[F(t)]_\alpha = [F]_\alpha(t) = \{f(t) : f \in [F]_\alpha\}. \tag{2.73}$$

Example 2.5. The mapping $F(x) = Ax$, where $A = [-1, 1]$ and $x \in \mathbb{R}$, is a set-valued function whose images are intervals.

Example 2.6. The mapping $F(x) = Ax$, where $A = (-1; 0; 1)$ and $x \in \mathbb{R}$, is a fuzzy-set-valued function whose images are triangular fuzzy numbers.

Example 2.7. Consider f_1, f_2, and f_3 continuous functions on an interval $I = [a, b]$. The fuzzy subset $F \in \mathscr{F}(C([a, b]; \mathbb{R}))$ such that

$$\mu_F(f) = \begin{cases} \alpha, & \text{if } f = f_1 + \alpha(f_2 - f_1) \\ \alpha, & \text{if } f = f_3 + \alpha(f_2 - f_3) \ , \qquad \alpha \in [0, 1], \\ 0, & \text{otherwise} \end{cases} \tag{2.74}$$

has triangular fuzzy numbers as attainable fuzzy sets. This is defined in [14] and is a particular kind of fuzzy bunch of functions, called *triangular fuzzy function*.

The fuzzy-number-valued function of the previous example can be constructed by considering the triangular fuzzy function with $f_1(x) = -x, f_2(x) = 0$ and $f_3(x) = x, x \in \mathbb{R}$ and calculating its attainable sets.

Using the definition of attainable sets, to each fuzzy bunch there corresponds only one fuzzy-set-valued function. But the converse is not true, as the next examples illustrate.

Example 2.8. The authors in [2] define the fuzzy bunches in $\mathscr{F}_{\mathscr{K}}(\mathscr{A}C([0, 2]; \mathbb{R}))$ [see Appendix A for definition of $\mathscr{A}C([a, b]; \mathbb{R})$] such that

$$F_1 = \{x(\cdot) : x(t) = a, a \in [0, 2]\}$$
$$F_2 = F_1 \cup \{y(\cdot) : y(t) = 2 - t\}$$

(2.75)

where $x, y : [0, 2] \to [0, 2]$.

We have $[F_1]_\alpha = [F_2]_\alpha = [0, 2]$, for all $\alpha \in [0, 1]$, though $F_1 \neq F_2$.

Example 2.9. The fuzzy bunches of functions $F_1, F_2 \in \mathscr{F}(C([-1, 1]; \mathbb{R}))$

$$\mu_{F_1}(f) = \begin{cases} \alpha, & \text{if } f : f(x) = -x(1 - \alpha) \\ \alpha, & \text{if } f : f(x) = x(1 - \alpha) \\ 0, & \text{otherwise} \end{cases} ,$$

(2.76)

and

$$\mu_{F_2}(f) = \begin{cases} \alpha, & \text{if } f : f(x) = -|x|(1 - \alpha) \\ \alpha, & \text{if } f : f(x) = |x|(1 - \alpha) \\ 0, & \text{otherwise} \end{cases} ,$$

(2.77)

are not equal, though their attainable sets are the same: they are the images of the function in Example 2.5. Each function in F_1 is a straight line on $[-1, 1]$, different from the functions in F_2, as Figs. 2.6 and 2.7 illustrate.

Fig. 2.6 A function of the support of the fuzzy bunch of functions F_1 of Example 2.9

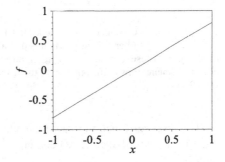

Fig. 2.7 A function of the support of the fuzzy bunch of functions F_2 of Example 2.9

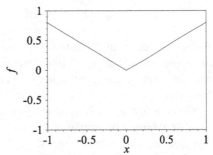

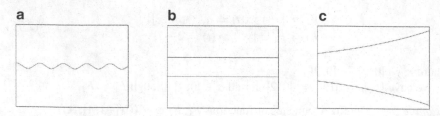

Fig. 2.8 Level set functions of the (**a**) 1-cut, (**b**) 0.6-cut, and (**c**) 0.2-cut of a fuzzy-number-valued function

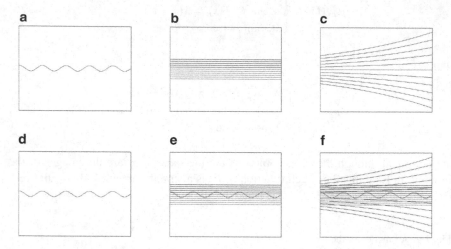

Fig. 2.9 Convex combinations of the level set functions of the (**a**) 1-cut, (**b**) 0.6-cut, and (**c**) 0.2-cut of a fuzzy-number-valued function and construction of the α-cuts of the representative bunch of first kind. These α-cuts of the representative bunch of first kind are defined as the union of the convex combinations corresponding to the α-cut and the α-cuts above: (**d**) 1-cut, (**e**) 0.6-cut, and (**f**) 0.2-cut

The level set function that defines the α-cuts of a fuzzy-number-valued function $F : x \mapsto F(x)$ will always be denoted $f_\alpha^-(x)$ and $f_\alpha^+(x)$ for this text. That is,

$$[F(x)]_\alpha = [f_\alpha^-(x), f_\alpha^+(x)]. \tag{2.78}$$

We are interested in defining fuzzy bunches of functions from fuzzy-number-valued functions such that the former preserves the main properties of the latter. *The main property is the equivalence of its attainable sets and the fuzzy-number-valued function.*

Given a fuzzy-number-valued function $F : [a, b] \rightarrow \mathscr{F}_\mathscr{C}(\mathbb{R})$, the idea (in order to a fuzzy bunch be similar in its properties to that of the fuzzy-set-valued function that generated it) is to consider the convex combinations of the functions f_α^- and f_α^+ (see Figs. 2.8a–c and 2.9a–c). This assures that the attainable sets of the α-cuts of the fuzzy bunches have no "holes," that is, they are convex subsets of $\mathbb{R}$. Convexity

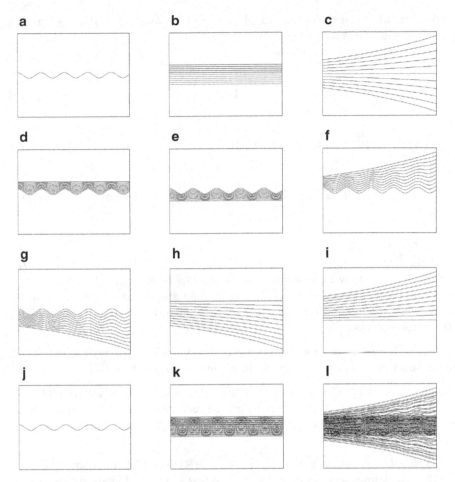

Fig. 2.10 Convex combinations of the level set functions of the (**a**) 1-cut, (**b**) 0.6-cut, (**c**) 0.2-cut, (**d**), (**e**) 1-cut with 0.6-cut, (**f**), (**g**) 1-cut with 0.2-cut, and (**h**), (**i**) 0.2-cut with 0.6-cut of a fuzzy-number-valued function and construction of the α-cuts of the representative bunch of second kind. These α-cuts of the representative bunch of second kind are defined as the union of the convex combinations corresponding to the α-cut and the α-cuts above: (**j**) 1-cut, (**k**) 0.6-cut, and (**l**) 0.2-cut

also preserves properties such as continuity and differentiability, which is a very important point. A possible problem is that an α-cut may not contain another α-cut with smaller value of α. The solution is to take all convex combinations of the upper α-cuts (see Fig. 2.9d–f). Finally, to make compactness (a desirable property) more likely to occur, the closedness of each α-cut is a property that we will require.

Based on these observations, two types of fuzzy bunches are created. The first one is simpler and the second kind has more elements (contains the elements in the first kind, see Fig. 2.10). The reason why we define both kinds of fuzzy bunches has to do with differentiability which will be explained in Sect. 3.2.2.

Definition 2.16. Consider $F : [a, b] \rightarrow \mathscr{F}_{\mathscr{C}}(\mathbb{R})$ where $f_\alpha^-(x)$ and $f_\alpha^+(x)$ are continuous functions with respect to x.

Define the subsets of functions

$$A_\alpha = \text{cl}\left(\bigcup_{\beta \geq \alpha} \bigcup_{0 \leq \lambda \leq 1} f_\beta^\lambda(\cdot)\right), \quad \alpha \in [0, 1] \tag{2.79}$$

where $f_\beta^\lambda(\cdot) = (1 - \lambda)f_\beta^-(\cdot) + \lambda f_\beta^+(\cdot)$ and

$$B_\alpha = \text{cl}\left(\bigcup_{\beta_1, \beta_2 \geq \alpha} \bigcup_{0 \leq \lambda \leq 1} f_{\beta_1, \beta_2}^\lambda(\cdot)\right), \quad \alpha \in [0, 1] \tag{2.80}$$

where $f_{\beta_1, \beta_2}^\lambda(\cdot) = (1 - \lambda)f_{\beta_1}^-(\cdot) + \lambda f_{\beta_2}^+(\cdot)$.

If the families $\{A_\alpha : \alpha \in [0, 1]\}$ and $\{B_\alpha : \alpha \in [0, 1]\}$ each define a fuzzy bunch of functions, we call them *representative affine fuzzy bunch of functions of first kind* (or representative bunch of first kind for short) and *representative affine fuzzy bunch of functions of second kind* (or representative bunch of second kind for short), respectively.

It is important to remark that whenever the symbol $\tilde{F}(x)$ is used where $\tilde{F}$ is a fuzzy bunch of functions, it refers to the attainable fuzzy sets of $\tilde{F}$ at x.

Example 2.10. Consider the triangular fuzzy functions of [14] (see Example 2.7) with $f_2 - f_1 = f_3 - f_2$, that is, the attainable sets are symmetrical triangular fuzzy numbers, and $f_1(x) \neq f_3(x)$ for all x. This is an example of representative bunches of first kind.

Example 2.11. A function $F : [a, b] \rightarrow \mathscr{F}_{\mathscr{C}}^0(\mathbb{R})$ [see (2.11)], where $f_\alpha^\pm(x)$ are continuous, defines representative bunches of first and second kinds in $C([a, b]; \mathbb{R})$.

Proof. In order to prove this statement, it suffices to demonstrate that the subsets A_α and B_α of Definition 2.16 satisfy conditions (i), (ii), (iii), and (iv) of Theorem 2.4. We prove this with respect to A_α. For B_α the reasoning is analogous.

Let us first prove (i), that is, A_α are nonempty compact sets, for all $\alpha \in [0, 1]$.

Since $A_\alpha = \text{cl}\left(\bigcup_{\beta \geq \alpha} \bigcup_{0 \leq \lambda \leq 1} f_\beta^\lambda(\cdot)\right)$ contains $f_\alpha^\pm(\cdot)$, it is nonempty. Note that the continuity in α implies

$$f_{\alpha_n}^\pm(x) \rightarrow f_\alpha^\pm(x) \quad \text{if} \quad \alpha_n \rightarrow \alpha \tag{2.81}$$

for $\alpha_n, \alpha \in I \subset [0, 1]$, for all $x \in [a, b]$. According to Dini's Theorem [3], pointwise convergence implies uniform convergence if the pointwise limits define a continuous function, the sequence of functions is monotonic and each function is defined on a compact set. Since this is the case, the convergence is uniform in x.

Hence

$$f_{\alpha_n}^{\pm}(\cdot) \to f_{\alpha}^{\pm}(\cdot) \quad \text{if} \quad \alpha_n \to \alpha. \tag{2.82}$$

Similarly,

$$f_{\alpha}^{\lambda_n}(\cdot) \to f_{\alpha}^{\lambda}(\cdot) \quad \text{if} \quad \lambda_n \to \lambda. \tag{2.83}$$

As a consequence,

$$f_{\alpha_n}^{\lambda_n}(\cdot) \to f_{\alpha}^{\lambda}(\cdot) \quad \text{if} \quad \alpha_n \to \alpha \text{ and } \lambda_n \to \lambda \tag{2.84}$$

for $\alpha_n, \alpha \in I \subset [0, 1]$ and $\lambda_n, \lambda \in [0, 1]$.

This means that $\bigcup_{\beta \geq \alpha} \bigcup_{0 \leq \lambda \leq 1} f_{\beta}^{\lambda}(\cdot)$ is sequentially compact. Since $C([a, b]; \mathbb{R})$ is a metric space, sequentially compactness is equivalent to compactness. Hence $\bigcup_{\beta \geq \alpha} \bigcup_{0 \leq \lambda \leq 1} f_{\beta}^{\lambda}(\cdot)$ is closed and equals A_α.

Condition (ii) states that if $0 \leq \alpha_1 \leq \alpha_2 \leq 1$ then $A_{\alpha_2} \subseteq A_{\alpha_1}$. Indeed,

$$A_{\alpha_2} = \bigcup_{\beta \geq \alpha_2} \bigcup_{0 \leq \lambda \leq 1} f_{\beta}^{\lambda}(\cdot) \subseteq \left(\bigcup_{\alpha_1 \leq \beta < \alpha_2} \bigcup_{0 \leq \lambda \leq 1} f_{\beta}^{\lambda}(\cdot) \right) \bigcup \left(\bigcup_{\beta \geq \alpha_2} \bigcup_{0 \leq \lambda \leq 1} f_{\beta}^{\lambda}(\cdot) \right) = A_{\alpha_1}. \tag{2.85}$$

We now prove condition (iii), that is, for any nondecreasing sequence (α_n) in $[0, 1]$ converging to $\alpha \in (0, 1]$ we have $\cap_{n=1}^{\infty} A_{\alpha_n} = A_\alpha$. From condition (ii) we have $A_\alpha \subseteq \cap_{n=1}^{\infty} A_{\alpha_n}$. To prove $A_\alpha \supseteq \cap_{n=1}^{\infty} A_{\alpha_n}$ consider $f \in \cap_{n=1}^{\infty} A_{\alpha_n}$. The function f is in each A_{α_n} and it can be written as $f = f_{\beta_n}^{\overline{\lambda}_n}$, with $\beta_n \in [\alpha_n, 1]$, $\overline{\lambda}_n \in [0, 1]$ (it is the same function f but written differently, according to the set A_α in which it is). Hence β_n admits subsequence converging to $\beta \in [\alpha, 1]$ and $\overline{\lambda}_n$ admits subsequence converging to $\overline{\lambda} \in [0, 1]$, so that it defines $f_{\beta}^{\overline{\lambda}} \in A_\alpha$. Therefore, $A_\alpha \supseteq \cap_{n=1}^{\infty} A_{\alpha_n}$ and condition (iii) is proved.

The last condition is proved if for any nonincreasing sequence (α_n) in $[0, 1]$ converging to zero we have $\text{cl}\left(\cup_{n=1}^{\infty} A_{\alpha_n} \right) \subseteq A_0$ and $\text{cl}\left(\cup_{n=1}^{\infty} A_{\alpha_n} \right) \supseteq A_0$. We first simplify the expression

$$\text{cl}\left(\bigcup_{n=1}^{\infty} A_{\alpha_n} \right) = \text{cl}\left(\bigcup_{n=1}^{\infty} \bigcup_{\beta \geq \alpha_n} \bigcup_{0 \leq \lambda \leq 1} f_{\beta}^{\lambda}(\cdot) \right) = \text{cl}\left(\bigcup_{\beta > 0} \bigcup_{0 \leq \lambda \leq 1} f_{\beta}^{\lambda}(\cdot) \right). \tag{2.86}$$

Note that

$$\text{cl}\left(\bigcup_{\beta > 0} \bigcup_{0 \leq \lambda \leq 1} f_{\beta}^{\lambda}(\cdot) \right) \subseteq \text{cl}\left(\bigcup_{\beta \geq 0} \bigcup_{0 \leq \lambda \leq 1} f_{\beta}^{\lambda}(\cdot) \right) = A_0 \tag{2.87}$$

which is the first inclusion. To prove the second one we need to prove that $f \in$
$A_0 = \bigcup\limits_{\beta \geq 0} \bigcup\limits_{0 \leq \lambda \leq 1} f_\beta^\lambda(\cdot)$ implies $f \in \mathrm{cl}\left(\bigcup\limits_{\beta > 0} \bigcup\limits_{0 \leq \lambda \leq 1} f_\beta^\lambda(\cdot)\right)$. There are two possibilities
for $f \in A_0$: (a) $f \in \bigcup\limits_{\beta > 0} \bigcup\limits_{0 \leq \lambda \leq 1} f_\beta^\lambda(\cdot)$ or (b) $f \in \bigcup\limits_{\beta = 0} \bigcup\limits_{0 \leq \lambda \leq 1} f_\beta^\lambda(\cdot) = \bigcup\limits_{0 \leq \lambda \leq 1} f_0^\lambda(\cdot)$. We only
need to prove case (b). We use the fact that $F(x)$ is a fuzzy number and therefore
satisfies

$$\mathrm{cl}\left(\bigcup_{n=1}^{\infty}[f_{\alpha_n}^-(x), f_{\alpha_n}^+(x)]\right) = [f_0^-(x), f_0^+(x)] \tag{2.88}$$

for (α_n) a nonincreasing sequence converging to zero. Hence

$$f_{\alpha_n}^\pm(x) \to f_0^\pm(x) \quad \text{and} \quad f_{\alpha_n}^{\lambda_n}(x) \to f_0^\lambda(x) \tag{2.89}$$

for $\alpha_n \searrow 0$ and $\lambda_n \to \lambda$, $\alpha_n \in [0, 1]$ and $\lambda_n, \lambda \in [0, 1]$. Using the same arguments
as before we have

$$f_{\alpha_n}^{\lambda_n}(\cdot) \to f_0^\lambda(\cdot) \tag{2.90}$$

uniformly and hence $f_0^\lambda(\cdot)$ is a point of closure. This means that

$$f_0^\lambda(\cdot) \in \mathrm{cl}\left(\bigcup_{\beta > 0} \bigcup_{0 \leq \lambda \leq 1} f_\beta^\lambda(\cdot)\right). \tag{2.91}$$

That is, the second inclusion is also satisfied and we have obtained the equality of
condition (iv).

Having proved (i), (ii), (iii), and (iv), it follows that A_α are α-cuts of the
representative bunch of first kind of F.

Example 2.12. Consider the fuzzy-number-valued function $F : [-1, 1] \to \mathscr{F}_\mathscr{C}(\mathbb{R})$
with α-cuts

$$[F(x)]_\alpha = \begin{cases} [10x^2 - 12, 10x^2 + 2], & \text{if } 0 \leq \alpha \leq 0.5 \\ [-1, 1], & \text{if } 0.5 < \alpha \leq 1 \end{cases}. \tag{2.92}$$

The representative bunch of first kind is given by the α-cuts

$$[\tilde{F}_1(\cdot)]_\alpha = \begin{cases} \bigcup\limits_{i=1}^{2} \bigcup\limits_{0 \leq \lambda \leq 1} y_i^\lambda(\cdot), & \text{if } 0 \leq \alpha \leq 0.5 \\ \bigcup\limits_{0 \leq \lambda \leq 1} y_1^\lambda(\cdot), & \text{if } 0.5 < \alpha \leq 1 \end{cases} \tag{2.93}$$

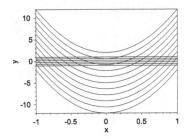

Fig. 2.11 Some elements of the support of the representative bunch of first kind of Example 2.12. The real-valued functions are the convex combinations of a level set function (of a fuzzy number-valued function) with the opposite level set function in the same α-cut

and the representative bunch of second kind is defined by

$$[\tilde{F}_1(\cdot)]_\alpha = \begin{cases} \bigcup\limits_{i=1}^{4} \bigcup\limits_{0 \leq \lambda \leq 1} y_i^\lambda(\cdot), \text{ if } 0 \leq \alpha \leq 0.5 \\ \bigcup\limits_{0 \leq \lambda \leq 1} y_1^\lambda(\cdot), \text{ if } 0.5 < \alpha \leq 1 \end{cases} \tag{2.94}$$

where

$$\begin{cases} y_1^\lambda(\cdot) : y_1^\lambda(x) = (1-\lambda)(10x^2 - 12) + \lambda(10x^2 + 2), \\ y_2^\lambda(\cdot) : y_2^\lambda(x) = (1-\lambda)(-1) + \lambda, \\ y_3^\lambda(\cdot) : y_3^\lambda(x) = (1-\lambda)(-1) + \lambda(10x^2 + 2), \\ y_4^\lambda(\cdot) : y_4^\lambda(x) = (1-\lambda)(10x^2 - 12) + \lambda, \end{cases} \tag{2.95}$$

for all $\lambda \in [0, 1]$.

Figures 2.11 and 2.12 present some of the elements of the supports of $\tilde{F}_1(\cdot)$ and $\tilde{F}_2(\cdot)$. It is remarkable that $\tilde{F}_2(\cdot)$ has more elements in its support than in the support of $\tilde{F}_1(\cdot)$ [in fact, $\tilde{F}_2(\cdot)$ contains $\tilde{F}_1(\cdot)$]. Moreover, some elements in $\tilde{F}_2(\cdot)$ have different behavior than those in $\tilde{F}_1(\cdot)$, though both fuzzy bunches have the same attainable sets.

Example 2.13. Consider the fuzzy-number-valued function $F : [0, 0.5] \rightarrow \mathscr{F}_\mathscr{C}^0(\mathbb{R})$ with α-cuts

$$[F(x)]_\alpha = \begin{cases} \left[x^2 - 3 + \alpha, (1 - 2\alpha)x^2 - 2\alpha + 2\right], \text{ if } 0 \leq \alpha \leq 0.5 \\ \left[x^2 - 3 + \alpha, (2\alpha - 1)x^2 - 6\alpha + 4\right], \text{ if } 0.5 < \alpha \leq 1 \end{cases}. \tag{2.96}$$

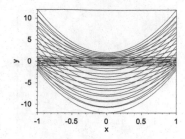

Fig. 2.12 Some elements of the support of the representative bunch of second kind of Example 2.12. The real-valued functions are the convex combinations of a level set function (of a fuzzy number-valued function) with opposite level set function that may belong to different α-cuts

The representative bunch of first kind is given by the α-cuts

$$
[\tilde{F}_1(\cdot)]_\alpha =
\begin{cases}
\text{cl}\left\{ \left(\displaystyle\bigcup_{\beta > 0.5} \bigcup_{0 \le \lambda \le 1} f_\beta^\lambda \right) \cup \left(\displaystyle\bigcup_{\alpha \le \beta \le 0.5} \bigcup_{0 \le \lambda \le 1} g_\beta^\lambda \right) \right\}, \text{ if } 0 \le \alpha \le 0.5 \\[3mm]
\qquad\qquad\qquad\qquad\qquad\qquad \displaystyle\bigcup_{\beta \ge \alpha} \bigcup_{0 \le \lambda \le 1} f_\beta^\lambda, \text{ if } 0.5 < \alpha \le 1
\end{cases}
$$

$$
=
\begin{cases}
\left(\displaystyle\bigcup_{\beta \ge 0.5} \bigcup_{0 \le \lambda \le 1} f_\beta^\lambda \right) \cup \left(\displaystyle\bigcup_{\alpha \le \beta \le 0.5} \bigcup_{0 \le \lambda \le 1} g_\beta^\lambda \right), \text{ if } 0 \le \alpha \le 0.5 \\[3mm]
\qquad\qquad\qquad\qquad\qquad \displaystyle\bigcup_{\beta \ge \alpha} \bigcup_{0 \le \lambda \le 1} f_\beta^\lambda, \text{ if } 0.5 < \alpha \le 1
\end{cases}
\tag{2.97}
$$

where

$$
\begin{cases}
f_\beta^\lambda(\cdot) : f_\beta^\lambda(x) = (1 - \lambda)(x^2 - 3 + \beta) + \lambda((2\beta - 1)x^2 - 6\beta + 4), \\
g_\beta^\lambda(\cdot) : g_\beta^\lambda(x) = (1 - \lambda)(x^2 - 3 + \beta) + \lambda((1 - 2\beta)x^2 - 2\beta + 2),
\end{cases}
\tag{2.98}
$$

for all $\lambda \in [0, 1]$ and $\beta \in [0, 1]$.

We define semicontinuity of set-valued functions as follows.

Definition 2.17 ([8]). A set-valued function $F : \Omega \to \mathscr{P}(\mathbb{R}^n)$, $\Omega \subset \mathbb{R}^m$, is *upper semicontinuous (usc)* at $t_0 \in \Omega$ if for every $\epsilon > 0$ there exists a $\delta = \delta(t_0, \epsilon) > 0$ such that

$$
\rho(F(t), F(t_0)) < \epsilon
\tag{2.99}
$$

if $||t - t_0|| < \delta$, $t \in \Omega$.

Definition 2.18 ([8]). A set-valued function $F : \Omega \to \mathscr{P}(\mathbb{R}^n)$, $\Omega \subset \mathbb{R}^m$, is *lower semicontinuous (lsc)* at $t_0 \in \Omega$ if for every $\epsilon > 0$ there exists a $\delta = \delta(t_0, \epsilon) > 0$ such that

$$\rho(F(t_0), F(t)) < \epsilon \qquad (2.100)$$

if $||t - t_0|| < \delta, t \in \Omega$.

The set-valued function F is said to be usc (lsc) if it is usc (lsc) at every $t \in \Omega$. If it is both usc and lsc, the function is continuous.

Example 2.14. The set-valued functions $F, G : \mathbb{R} \to \mathscr{P}(\mathbb{R})$ such that

$$F(x) = \begin{cases} [-1, 1], & \text{if } x = 0 \\ \{0\}, & \text{if } x \neq 0 \end{cases} \quad \text{and} \quad G(x) = \begin{cases} [-1, 1], & \text{if } x \neq 0 \\ \{0\}, & \text{if } x = 0 \end{cases} \qquad (2.101)$$

are usc and lsc, respectively.

The concept of semicontinuity of fuzzy-set-valued functions is similar.

Definition 2.19 ([8]). A fuzzy-set-valued function $F : \Omega \to \mathscr{F}_{\mathscr{K}}(\mathbb{R}^n)$, $\Omega \in \mathbb{R}^m$, is *upper semicontinuous (usc)* at $t_0 \in \Omega$ if for every $\epsilon > 0$ there exists a $\delta = \delta(t_0, \epsilon) > 0$ such that

$$\rho([F(t_0)]^\alpha, [F(t)]^\alpha) < \epsilon \qquad (2.102)$$

if $||t - t_0|| < \delta, t \in \Omega$, for all $\alpha \in [0, 1]$.

Definition 2.20 ([8]). A fuzzy-set-valued function $F : \Omega \to \mathscr{F}_{\mathscr{K}}(\mathbb{R}^n)$, $\Omega \in \mathbb{R}^m$, is *lower semicontinuous (lsc)* at $t_0 \in \Omega$ if for every $\epsilon > 0$ there exists a $\delta = \delta(t_0, \epsilon) > 0$ such that

$$\rho([F(t)]^\alpha, [F(t_0)]^\alpha) < \epsilon \qquad (2.103)$$

if $||t - t_0|| < \delta, t \in \Omega$, for all $\alpha \in [0, 1]$.

The fuzzy-set-valued function F is said to be usc (lsc) if it is usc (lsc) at every $t \in \Omega$. If the fuzzy-set-valued function F is usc (lsc), the set-valued functions $[F]^\alpha : \Omega \to \mathscr{K}^n$ are clearly usc (lsc). The converse implication is not necessarily true, unless $[F]^\alpha$ are uniformly usc (lsc) in $\alpha \in [0, 1]$. As a result of these definitions, a fuzzy-set-valued function F is d_∞-continuous if and only if it is usc and lsc. In this text if a function is d_∞-continuous it will be said that it is continuous. If another metric is used, we will specify it.

Denote by $C([a, b]; \mathscr{F}_{\mathscr{C}}(\mathbb{R}^n))$ the space of continuous fuzzy-set-valued functions from $[a, b]$ to $\mathscr{F}_{\mathscr{C}}(\mathbb{R}^n)$ endowed with the metric $H(F, G) = \sup_{x \in [a,b]} d_\infty(F(x), G(x))$ for $F, G \in C([a, b]; \mathscr{F}_{\mathscr{C}}(\mathbb{R}^n))$. The next result is important to assure existence of solution of FDEs.

Theorem 2.10 ([4]). *The space $C([a, b]; \mathscr{F}_{\mathscr{C}}(\mathbb{R}))$ is a complete metric space.*

Summary of fuzzy functions:

- **Two kinds.** This text explores two kinds of fuzzy functions: the fuzzy-number-valued functions (from $\mathbb{R}$ to the space of fuzzy numbers $\mathscr{F}_{\mathscr{C}}(\mathbb{R})$) and fuzzy bunches of functions (fuzzy subsets of spaces of functions).
- **One fuzzy-number-valued function to one fuzzy bunch of functions.** Each fuzzy bunch of functions may define a fuzzy-number-valued function via attainable sets.
- **Various fuzzy bunches of functions to one fuzzy-number-valued function.** Each fuzzy-number-valued function may define several fuzzy bunches of functions. The definition of the representative bunches of first and second kind help us to choose one fuzzy bunch.

2.6 Summary

Some basic concepts and results were presented in this chapter and we briefly summarize them:

- A fuzzy set is a set characterized by a membership function $\mu_A : \mathbb{U} \to [0, 1]$, where $\mu_A(x) = 1$ means x is completely in the subset A, $\mu_A(x) = 0$ means x is completely out the subset A and $0 < \mu_A(x) < 1$ belongs to A with an intermediate degree.
- The α-cut of a fuzzy subset in $\mathbb{U}$ is the collection of elements whose membership function is at least α.
- A fuzzy number is a particular kind of fuzzy subset in $\mathbb{R}$.
- The α-cuts of a fuzzy number are nonempty compact intervals. Also, a family of nonempty compact intervals satisfying some conditions always defines a fuzzy number.
- Fuzzy-set-valued functions are functions corresponding real numbers to fuzzy subsets.
- Fuzzy bunches of functions are fuzzy subsets of a space of functions.
- To relate these two last concepts we use the attainable fuzzy sets (from a fuzzy bunch we obtain a fuzzy-set-valued function) and the representative fuzzy bunches (from a fuzzy-set-valued function we obtain a fuzzy bunch).

References

1. L.C. Barros, R.C. Bassanezi, P.A. Tonelli, On the continuity of Zadeh's extension, in *Proceedings of the Seventh IFSA World Congress*, vol. 2 (1997), pp. 3–8
2. L.C. Barros, P.A. Tonelli, A.P. Julião, Cálculo diferencial e integral para funções fuzzy via extensão dos operadores de derivação e integração. Technical Report 6 (2010) [in Portuguese]
3. R.G. Bartle, D.R. Sherbert, *Introduction to Real Analysis* (Wiley, Berlin, 2011)
4. B. Bede, *Mathematics of Fuzzy Sets and Fuzzy Logic* (Springer, Berlin/Heidelberg, 2013)
5. B. Bede, L. Stefanini, Generalized differentiability of fuzzy-valued functions. Fuzzy Sets Syst. **230**, 119–141 (2013)
6. C. Carlsson, R. Fullér, P. Majlender, Additions of completely correlated fuzzy numbers, in *IEEE International Conference on Fuzzy Systems* (2004), pp. 535–539
7. M.S. Cecconello, *Sistemas Dinâmicos em Espaços Métricos Fuzzy – Aplicações em Biomatemática*. PhD thesis, IMECC – UNICAMP (2010) [in Portuguese]
8. P. Diamond, P. Kloeden, *Metric Spaces of Fuzzy Sets: Theory and Applications* (World Scientific, Singapore, 1994)
9. D. Dubois, H. Prade, *Fuzzy Sets and Systems: Theory and Applications* (Academic, Orlando, 1980)
10. D. Dubois, H. Prade, Additions of interactive fuzzy numbers. IEEE Trans. Autom. Control **26**, 926–936 (1981)
11. D. Dubois, H. Prade, Gradual elements in a fuzzy set. Soft Comput. **12**, 165–175 (2008)
12. R. Fullér, *Fuzzy Reasoning and Fuzzy Optimization* (Turku Centre for Computer Science, Abo, 1998)
13. R. Fullér, T. Keresztfalvi, t-norm-based addition of fuzzy intervals. Fuzzy Sets Syst. **51**, 155–159 (1992)
14. N.A. Gasilov, I.F. Hashimoglu, S.E. Amrahov, A.G. Fatullayev, A new approach to non-homogeneous fuzzy initial value problem. Comput. Model. Eng. Sci. **85**, 367–378 (2012)
15. L.T. Gomes, L.C. Barros, A note on the generalized difference and the generalized differentiability. Fuzzy Sets Syst. (2015). doi:10.1016/j.fss.2015.02.015
16. H. Huang, C. Wu, Approximation of fuzzy functions by regular fuzzy neural networks. Fuzzy Sets Syst. **177**, 60–79 (2011)
17. M. Hukuhara, Intégration des applications measurables dont la valeur est un compact convexe. Funkc. Ekvacioj **10**, 205–223 (1967)
18. O. Kaleva, The Cauchy problem for fuzzy differential equations. Fuzzy Sets Syst. **35**, 389–396 (1990)
19. G.J. Klir, B. Yuan, *Fuzzy Sets and Fuzzy Logic: Theory and Applications* (Prentice Hall, Upper Saddle River, 1995)
20. P.E. Kloeden, Compact supported endographs and fuzzy sets. Fuzzy Sets Syst. **4**, 193–201 (1980)
21. W.A. Lodwick, Constrained interval arithmetic. Technical Report CCM Report 138 (February 1999)
22. W.A. Lodwick, Fundamentals of interval analysis and linkages to fuzzy sets theory, in *Handbook of Granular Computing*, chapter 13, ed. by W. Pedrycz, A. Skowron, V. Kreinovich (Wiley, Chichester, 2008)
23. W.A. Lodwick, O.A. Jenkins, Constrained intervals and interval spaces. Soft Comput. **17**, 1393–1402 (2013)
24. W.A. Lodwick, E.A. Untiedt, A comparison of interval analysis using constraint interval arithmetic and fuzzy interval analysis using gradual numbers, in *Annual Meeting of the North American Fuzzy Information Processing Society* (2008), pp. 1–6
25. M. Ma, On embedding problems of fuzzy number space: Part 4. Fuzzy Sets Syst. **58**, 185–193 (1993)
26. M. Mizumoto, K. Tanaka, The four operations of arithmetic on fuzzy numbers. Syst. Comput. Control **7**, 73–81 (1976)

27. R.E. Moore, *Interval Analysis* (Prentice-Hall, New York, 1966)
28. C.V. Negoita, D.A. Ralescu, *Applications of Fuzzy Sets to Systems Analysis* (Wiley, New York, 1975)
29. H.T. Nguyen, A note on the extension principle for fuzzy sets. J. Math. Anal. Appl. **64**, 369–380 (1978)
30. H.T. Nguyen, E.A. Walker, *A First Course in Fuzzy Logic* (Taylor & Francis, Boca Raton, 2005)
31. W. Pedrycz, F. Gomide, *Fuzzy Systems Engineering – Toward Human Centric Computing*. (Wiley & Sons, New Jersey, 2007)
32. M. Puri, D. Ralescu, Differentials of fuzzy functions. J. Math. Anal. Appl. **91**, 552–558 (1983)
33. M. Puri, D. Ralescu, Fuzzy random variables. J. Math. Anal. Appl. **114**, 409–422 (1986)
34. L. Stefanini, A generalization of Hukuhara difference and division for interval and fuzzy arithmetic. Fuzzy Sets Syst. **161**, 1564–1584 (2010)
35. L. Stefanini, B. Bede, Generalized Hukuhara differentiability of interval-valued functions and interval differential equations. Nonlinear Anal. Theory Methods Appl. **71**, 1311–1328 (2009)
36. L.A. Zadeh, The concept of linguistic variable and its application to approximate reasoning – I. Inf. Sci. **8**, 199–249 (1975)

Chapter 3
Fuzzy Calculus

This chapter treats two types of fuzzy calculus: one for fuzzy-set-valued functions and other for fuzzy bunches of functions. Section 3.1 reviews definitions of fuzzy Aumann, Henstock, and Riemann integrals and the Hukuhara derivative and its generalizations. It also provides some theorems, including a Fundamental Theorem of Calculus. All these definitions and results were previously presented in the literature. Section 3.2 introduces derivative and integral for fuzzy bunches of functions and results concerning them, some of which never published before. Examples illustrate some of the concepts and theorems, especially in the last section, where new results provide comparisons between the different approaches.

3.1 Fuzzy Calculus for Fuzzy-Set-Valued Functions

This section reviews some known approaches of integrals (Aumann, Riemann, and Henstock integrals) and derivatives (Hukuhara and generalized derivatives) for fuzzy-set-valued functions. It also presents results connecting these fuzzy integrals and derivatives. The reader interested in other proposals may refer to (e.g., [11–13, 16]).

3.1.1 Integrals

The first integral proposed for fuzzy-number-valued functions is based on Aumann integral for multivalued functions [2] and was defined in [21] and [23].

© The Author(s) 2015
L.T. Gomes et al., *Fuzzy Differential Equations in Various Approaches*,
SpringerBriefs in Mathematics, DOI 10.1007/978-3-319-22575-3_3

Denote by $S(G)$ the subset of all integrable selections of a set-valued function $G : I \to \mathscr{P}(\mathbb{R}^n)$, i.e.,

$$S(G) = \{g : I \to \mathbb{R}^n : g \text{ is integrable and } g(t) \in G(t), \forall t \in I\}. \tag{3.1}$$

Definition 3.1 ([21, 23]). The Aumann integral of a fuzzy-set-valued function $F : [a, b] \to \mathscr{F}_{\mathscr{C}}(\mathbb{R}^n)$ over $[a, b]$ is defined levelwise

$$\left[(A) \int_a^b F(x) \, dx \right]_\alpha = \int_a^b [F]_\alpha \, dx \tag{3.2}$$

$$= \left\{ \int_a^b g(x) \, dx : g \in S([F(x)]_\alpha) \right\} \tag{3.3}$$

for all $\alpha \in [0, 1]$.

The function $F : [a, b] \to \mathscr{F}_{\mathscr{C}}(\mathbb{R}^n)$ is said to be Aumann integrable over $[a, b]$ if $(A) \int_a^b F(x) \, dx \in \mathscr{F}_{\mathscr{C}}(\mathbb{R}^n)$.

The following integrals have been defined for functions $F : [a, b] \to \mathscr{F}_{\mathscr{C}}(\mathbb{R})$.

Definition 3.2 ([15, 26]). The Riemann integral of a fuzzy-number-valued function $F : [a, b] \to \mathscr{F}_{\mathscr{C}}(\mathbb{R})$ over $[a, b]$ is the fuzzy number A such that for every $\epsilon > 0$ there exist $\delta > 0$ such that for any division $d : a = x_0 < x_1 < \ldots < x_n = b$ with $x_i - x_{i-1} < \delta, i = 1, \ldots, n$, and $\xi_i \in [x_i - x_{i-1}]$

$$d_\infty \left(\sum_{i=1}^{n-1} F(\xi_i)(x_i - x_{i-1}), A \right) < \epsilon. \tag{3.4}$$

The function $F : [a, b] \to \mathscr{F}_{\mathscr{C}}(\mathbb{R})$ is said to be Riemann integrable over $[a, b]$ if $A \in \mathscr{F}_{\mathscr{C}}(\mathbb{R})$. We denote $(R) \int_a^b F(x) \, dx = A$

Definition 3.3 ([7, 26]). Consider $\delta_n : a = x_0 < x_1 < \ldots < x_n = b$ a partition of the interval $[a, b]$, $\xi_i \in [x_i - x_{i-1}]$, $i = 1, \ldots, n$, a sequence ξ in δ_n and $\delta(x) > 0$ a real-valued function over $[a, b]$. The division $P(\delta_n, \xi)$ is considered to be δ-fine if

$$[x_{i-1}, x_i] \subseteq (\xi_{i-1} - \delta(\xi_{i-1}), \xi_{i-1} + \delta(\xi_{i-1})) \tag{3.5}$$

The Henstock integral of a fuzzy-number-valued function $F : [a, b] \to \mathscr{F}_{\mathscr{C}}(\mathbb{R})$ over $[a, b]$ is the fuzzy number A such that for every $\epsilon > 0$ there exist a real-valued function δ such that for any δ-fine division $P(\delta_n, \xi)$,

$$d_\infty \left(\sum_{i=1}^{n-1} F(\xi_i)(x_i - x_{i-1}), A \right) < \epsilon. \tag{3.6}$$

The function $F : [a, b] \rightarrow \mathscr{F}_{\mathscr{C}}(\mathbb{R})$ is said to be Henstock integrable over $[a, b]$ if $A \in \mathscr{F}_{\mathscr{C}}(\mathbb{R})$. We denote $(H) \int_a^b F(x)\, dx = A$.

Henstock integral is more general than Riemann, i.e., whenever a function is Riemann integrable, it is Henstock integrable as well.

Remark 3.1. Writing that a function is *integrable*, without specifying whether it is Aumann, Riemann, or Henstock, means it is integrable in all these three senses.

Corollary 3.1 ([5, 21, 26]). *If a function $F : [a, b] \rightarrow \mathscr{F}_{\mathscr{C}}(\mathbb{R})$ is continuous, then it is integrable. Moreover,*

$$\left[\int F \right]_\alpha = \left[\int f_\alpha^-, \int f_\alpha^+ \right] \tag{3.7}$$

for all $\alpha \in [0, 1]$.

Theorem 3.1 ([5, 21, 26]). *Let $F : [a, b] \rightarrow \mathscr{F}_{\mathscr{C}}(\mathbb{R})$ be integrable and $a \le x_1 \le x_2 \le x_3 \le b$. Then*

$$\int_{x_1}^{x_3} F = \int_{x_1}^{x_2} F + \int_{x_2}^{x_3} F. \tag{3.8}$$

Theorem 3.2 ([5, 21, 26]). *Let $F, G : [a, b] \rightarrow \mathscr{F}_{\mathscr{C}}(\mathbb{R})$ be integrable, then*

(i) $\int (F + G) = \int F + \int G$;
(ii) $\int (\lambda F) = \lambda \int F$, for any $\lambda \in \mathbb{R}$;
(iii) $d_\infty(F, G)$ is integrable;
(iv) $d_\infty(\int F, \int G) \le \int d_\infty(F, G)$.

3.1.2 Derivatives

The Hukuhara differentiability for fuzzy functions is based on the concept of Hukuhara differentiability for interval-valued functions [20].

Definition 3.4 ([22]). Let $F : (a, b) \rightarrow \mathscr{F}_{\mathscr{C}}(\mathbb{R}^n)$. If the limits

$$\lim_{h \to 0^+} \frac{F(x_0 + h) \ominus_H F(x_0)}{h} \quad \text{and} \quad \lim_{h \to 0^+} \frac{F(x_0) \ominus_H F(x_0 - h)}{h} \tag{3.9}$$

exist and equal some element $F'_H(x_0) \in \mathscr{F}_{\mathscr{C}}(\mathbb{R}^n)$, then F is Hukuhara differentiable (H-differentiable for short) at x_0 and $F'_H(x_0)$ is its Hukuhara derivative (H-derivative for short) at x_0.

Example 3.1. The fuzzy-number-valued function of Example 2.6, $F(x) = Ax$ with $A = (-1; 0; 1)$, is an H-differentiable function for $x \ge 0$ and

$$F'_H(x) = A. \tag{3.10}$$

For $x < 0$, F is not H-differentiable since $F(x + h) \ominus_H F(x)$ is not defined. Considering $x > 0$, F is a particular case of Example 8.30 in [5], which shows that any function $G(x) = Bg(x)$ with $g(x) > 0$, $g'(x) > 0$ and B a fuzzy number is H-differentiable. Moreover,

$$G'_H(x) = Bg'(x). \tag{3.11}$$

An H-differentiable fuzzy function has H-differentiable α-cuts (that is, its α-cuts are interval-valued H-differentiable functions). The converse, however, is not true, unless its α-cuts are uniformly H-differentiable (see [21]).

Definition 3.5 ([24]). Let $F : [a, b] \to \mathscr{F}_{\mathscr{C}}(\mathbb{R})$. If

$$[(f_\alpha^-)'(x_0), (f_\alpha^+)'(x_0)] \tag{3.12}$$

exists for all $\alpha \in [0, 1]$ and defines the α-cuts of a fuzzy number $F'_S(x_0)$, then F is Seikkala differentiable at x_0 and $F'_S(x_0)$ is the Seikkala derivative of F at x_0.

If $F : [a, b] \to \mathscr{F}_{\mathscr{C}}(\mathbb{R})$ is H-differentiable, then $f_\alpha^-(x)$ and $f_\alpha^+(x)$ are differentiable and

$$[F'(x_0)]_\alpha = [(f_\alpha^-)'(x_0), (f_\alpha^+)'(x_0)], \tag{3.13}$$

that is, if F is H-differentiable, it is Seikkala differentiable and the derivatives are the same [21].

Theorem 3.3 ([21]). *Let $F : [a, b] \to \mathscr{F}_{\mathscr{C}}(\mathbb{R}^n)$ be an H-differentiable function. Then it is continuous.*

Theorem 3.4 ([21]). *Let $F, G : [a, b] \to \mathscr{F}_{\mathscr{C}}(\mathbb{R}^n)$ be H-differentiable functions and $\lambda \in \mathbb{R}$. Then $(F + G)'_H = F'_H + G'_H$ and $(\lambda F)'_H = \lambda F'_H$.*

If F is Seikkala (or Hukuhara) differentiable, $(f_\alpha^-)'(x) \le (f_\alpha^+)'(x)$, hence the function diam $[F(x)]_\alpha = f_\alpha^+(x) - f_\alpha^-(x)$ is nondecreasing on $[a, b]$. It means that the function has nondecreasing fuzziness. As will be clear in Chap. 4, this is considered a shortcoming since an H-differentiable function cannot represent a function with decreasing fuzziness or periodicity. In order to overcome this, the generalized differentiability concepts were created. They generalize the H-differentiability, that is, they are defined for more cases of fuzzy-number-valued functions and whenever the H-derivative of a function exists, its generalization exists and has the same value.

Definition 3.6 ([6, 8]). Let $F : (a, b) \to \mathscr{F}_{\mathscr{C}}(\mathbb{R})$. If the limits of some pair

(i) $\displaystyle\lim_{h \to 0^+} \frac{F(x_0 + h) \ominus_H F(x_0)}{h}$ and $\displaystyle\lim_{h \to 0^+} \frac{F(x_0) \ominus_H F(x_0 - h)}{h}$ or

(ii) $\displaystyle\lim_{h \to 0^+} \frac{F(x_0) \ominus_H F(x_0 + h)}{-h}$ and $\displaystyle\lim_{h \to 0^+} \frac{F(x_0 - h) \ominus_H F(x_0)}{-h}$ or

(iii) $\lim\limits_{h\to 0^+} \dfrac{F(x_0 + h) \ominus_H F(x_0)}{h}$ and $\lim\limits_{h\to 0^+} \dfrac{F(x_0 - h) \ominus_H F(x_0)}{-h}$ or

(iv) $\lim\limits_{h\to 0^+} \dfrac{F(x_0) \ominus_H F(x_0 + h)}{-h}$ and $\lim\limits_{h\to 0^+} \dfrac{F(x_0) \ominus_H F(x_0 - h)}{h}$

exist and are equal to some element $F'_G(x_0)$ of $\mathscr{F}_\mathscr{C}(\mathbb{R})$, then F is strongly generalized differentiable (or GH-differentiable) at x_0 and $F'_G(x_0)$ is the strongly generalized derivative (GH-derivative for short) of F at x_0.

An (i)-strongly generalized differentiable function presents nondecreasing diameter, since it is the definition of the H-differentiability. (ii)-strongly generalized differentiability (we call (ii)-differentiability, for short), on the other hand, implies in nonincreasing diameter. The (iii) and (iv)-differentiability cases correspond to points where the function changes its behavior with respect to the diameter. It means that a strongly differentiable non-crisp function may present periodical behavior, as well as convergence to a single point.

In case F is defined on a closed interval, that is, $F : [a, b] \to \mathscr{F}_\mathscr{C}(\mathbb{R})$, we define the derivative at a using the limit from the right and at b using the limit from the left.

Example 3.2. The fuzzy-number-valued function of Example 2.6, $F(x) = Ax$ with $A = (-1; 0; 1)$, is a GH-differentiable function for $x \in \mathbb{R}$ and

$$F'_{gH}(x) = A. \tag{3.14}$$

Different from the H-derivative case, the GH-derivative of F is defined for $x < 0$. According to Example 8.35 in [5], any function $G(x) = Bg(x)$ with B a fuzzy number and $g : (a, b) \to \mathbb{R}$ differentiable with at most a finite number of roots in (a, b) is GH-differentiable. Moreover,

$$G'_H(x) = Bg'(x). \tag{3.15}$$

Example 3.2 illustrates that, different from the H-derivative, GH-differentiable functions can have decreasing diameter.

Definition 3.7 ([8]). Let $F : (a, b) \to \mathscr{F}_\mathscr{C}(\mathbb{R})$ and $x_0 \in (a, b)$. For a nonincreasing sequence $h_n \to 0$ and $n_0 \in \mathbb{N}$ we denote

$$A^{(1)}_{n_0} = \left\{ n \geq n_0; \exists E^{(1)}_n := F(x_0 + h_n) \ominus_H F(x_0) \right\}, \tag{3.16}$$

$$A^{(2)}_{n_0} = \left\{ n \geq n_0; \exists E^{(2)}_n := F(x_0) \ominus_H F(x_0 + h_n) \right\}, \tag{3.17}$$

$$A^{(3)}_{n_0} = \left\{ n \geq n_0; \exists E^{(3)}_n := F(x_0) \ominus_H F(x_0 - h_n) \right\}, \tag{3.18}$$

$$A^{(4)}_{n_0} = \left\{ n \geq n_0; \exists E^{(4)}_n := F(x_0 - h_n) \ominus_H F(x_0) \right\}. \tag{3.19}$$

The function F is said to be weakly generalized differentiable at x_0 if for any nonincreasing sequence $h_n \to 0$ there exists $n_0 \in \mathbb{N}$, such that

$$A_{n_0}^{(1)} \cup A_{n_0}^{(2)} \cup A_{n_0}^{(3)} \cup A_{n_0}^{(4)} = \{n \in \mathbb{N}; n \geq n_0\} \tag{3.20}$$

and moreover, there exists an element in $\mathscr{F}_C(\mathbb{R})$, such that if for some $j \in \{1, 2, 3, 4\}$ we have card $(A_{n_0}^{(j)}) = +\infty$, then

$$\lim_{h_n \searrow 0, n \to \infty, n \in A_{n_0}^{(j)}} d_\infty \left(\frac{E_n^{(j)}}{(-1)^{j+1} h_n}, F'(x_0) \right) = 0. \tag{3.21}$$

Definition 3.7 is more general than Definition 3.6, that is, it is defined for more cases of fuzzy-number-valued functions and whenever the latter exists, the former also exists and has the same value.

The next definition is equivalent to Definition 3.7 (see [10]).

Definition 3.8 ([10, 25]). Let $F : (a, b) \to \mathscr{F}_{\mathscr{C}}(\mathbb{R})$. If the limit

$$\lim_{h \to 0} \frac{F(x_0 + h) \ominus_{gH} F(x_0)}{h} \tag{3.22}$$

exists and belongs to $\mathscr{F}_{\mathscr{C}}(\mathbb{R})$, then F is generalized Hukuhara differentiable (gH-differentiable for short) at x_0 and $F'_{gH}(x_0)$ is the generalized Hukuhara derivative (gH-derivative for short) of F at x_0.

Theorem 3.5 ([10]). *Let $F : [a, b] \to \mathscr{F}_{\mathscr{C}}(\mathbb{R}^n)$ be a gH-differentiable function at x_0. Then it is levelwise continuous at x_0.*

Theorem 3.6 ([10]). *Let $F : [a, b] \to \mathscr{F}_{\mathscr{C}}(\mathbb{R})$ be such that the functions $f_\alpha^-(x)$ and $f_\alpha^+(x)$ are real-valued functions, differentiable with respect to x, uniformly in $\alpha \in [0, 1]$. Then the function $F(x)$ is gH-differentiable at a fixed $x \in [a, b]$ if and only if one of the following two cases holds:*

(a) $\left(f_\alpha^-\right)'(x)$ *is increasing,* $\left(f_\alpha^+\right)'(x)$ *is decreasing as functions of α, and* $\left(f_1^-\right)'(x) \leq \left(f_1^+\right)'(x)$, *or*

(b) $\left(f_\alpha^-\right)'(x)$ *is decreasing,* $\left(f_\alpha^+\right)'(x)$ *is increasing as functions of α, and* $\left(f_1^+\right)'(x) \leq \left(f_1^-\right)'(x)$.

Moreover,

$$\left[F'_{gH}(x)\right]_\alpha = [\min\{\left(f_\alpha^-\right)'(x), \left(f_\alpha^+\right)'(x)\}, \max\{\left(f_\alpha^-\right)'(x), \left(f_\alpha^+\right)'(x)\}], \tag{3.23}$$

for all $\alpha \in [0, 1]$.

The next concept further extends the gH-differentiability.

Definition 3.9 ([25]). Let $F : (a, b) \to \mathscr{F}_{\mathscr{C}}(\mathbb{R})$. If the limit

$$\lim_{h \to 0} \frac{F(x_0 + h) \ominus_g F(x_0)}{h} \tag{3.24}$$

exists and belongs to $\mathscr{F}_{\mathscr{C}}(\mathbb{R})$, then F is generalized differentiable (g-differentiable for short) at x_0 and $F'_g(x_0)$ is the fuzzy generalized derivative (g-derivative for short) of F at x_0.

Example 3.3. Recall the fuzzy-number-valued function of Example 2.13, F : $[0, 0.5] \rightarrow \mathscr{F}_{\mathscr{C}}(\mathbb{R})$ with α-cuts

$$[F(x)]_\alpha = \begin{cases} \left[x^2 - 3 + \alpha, (1 - 2\alpha)x^2 - 2\alpha + 2 \right], & \text{if } 0 \le \alpha \le 0.5 \\ \left[x^2 - 3 + \alpha, (2\alpha - 1)x^2 - 6\alpha + 4 \right], & \text{if } 0.5 < \alpha \le 1 \end{cases}. \quad (3.25)$$

The aim is to calculate the gH and the g-derivative of F.

Equation (2.53) provides easy means to calculate (3.22). For $\alpha \in [0, 0.5]$ one obtains

$$[F(x + h) \ominus_{gH} F(x)]_\alpha = [(1 - 2\alpha)(2xh + h^2), 2xh + h^2]. \quad (3.26)$$

Thus

$$\lim_{h \to 0} \frac{[F(x + h) \ominus_{gH} F(x)]_\alpha}{h} = [(1 - 2\alpha)2x, 2x] \quad (3.27)$$

and as consequence

$$\lim_{h \to 0} \frac{[F(x + h) \ominus_{gH} F(x)]_0}{h} = \{2x\} \quad (3.28)$$

and

$$\lim_{h \to 0} \frac{[F(x + h) \ominus_{gH} F(x)]_{0.25}}{h} = [x, 2x]. \quad (3.29)$$

The condition

$$\alpha < \beta \implies \lim_{h \to 0} \frac{[F(x + h) \ominus_{gH} F(x)]_\beta}{h} \subset \lim_{h \to 0} \frac{[F(x + h) \ominus_{gH} F(x)]_\alpha}{h} \quad (3.30)$$

does not hold, hence $\lim_{h \to 0} \frac{[F(x+h) \ominus_{gH} F(x)]_\alpha}{h}$ cannot be a fuzzy number and the gH-derivative is not defined for this function.

Equation (2.54) can be used in this case to find (3.24), for all $\alpha \in [0, 1]$. Since $f_\beta^-(x+h) - f_\beta^-(x) = 2xh + h^2$ and $f_\beta^+(x+h) - f_\beta^+(x) = (1 - 2\alpha)2xh + h^2$ for $\beta \le 0.5$ and $f_\beta^-(x+h) - f_\beta^-(x) = 2xh + h^2$ and $f_\beta^+(x+h) - f_\beta^+(x) = (2\alpha - 1)2xh + h^2$ for $\beta > 0.5$, we obtain for $\alpha > 0.5$:

$$\lim_{h \to 0} \frac{[F(x+h) \ominus_g F(x)]_\alpha}{h} = \mathrm{cl} \bigcup_{\beta \geq \alpha > 0.5} [(2\beta - 1)2x, 2x] = [(2\alpha - 1)2x, 2x].$$

(3.31)

For $\alpha \leq 0.5$, the levelwise limit becomes

$$\mathrm{cl} \left(\bigcup_{0.5 \geq \beta \geq \alpha \geq 0} [(1 - 2\beta)2x, 2x] \right) \bigcup \left(\bigcup_{\beta > 0.5} [(2\beta - 1)2x, 2x] \right) = [0, 2x]. \quad (3.32)$$

The result is the fuzzy number $F'_g : [0, 0.5] \to \mathscr{F}_{\mathscr{C}}(\mathbb{R})$ with α-cuts

$$[F'_g(x)]_\alpha = \begin{cases} [0, 2x], & \text{if } 0 \leq \alpha \leq 0.5 \\ [(2\alpha - 1)2x, 2x], & \text{if } 0.5 < \alpha \leq 1 \end{cases}. \quad (3.33)$$

as the g-derivative.

The g-difference is not defined for all pairs of fuzzy numbers, as we showed in Example 2.3. The same happens to the g-derivative, that is, it is not always well-defined (see also [17]).

Example 3.4. The definition of the g-derivative of the fuzzy-number-valued function of Example 2.12 leads to

$$[F'_g(x)]_\alpha = \begin{cases} \{20x\} \bigcup \{0\}, & \text{if } 0 \leq \alpha \leq 0.5 \\ \{0\}, & \text{if } 0.5 < \alpha \leq 1 \end{cases}. \quad (3.34)$$

That is, it is not a fuzzy-number-valued function. Hence F is not g-differentiable.

The function F in Example 3.4 has $f_\alpha^-(x)$ and $f_\alpha^+(x)$ differentiable real-valued functions with respect to x, uniformly with respect to $\alpha \in [0, 1]$, but it is not g-differentiable. In the case a function is g-differentiable and satisfy the just mentioned hypothesis, it has a formula that has been proved by [10].

Theorem 3.7. *Let $F : [a, b] \to \mathbb{R}_{\mathscr{F}}$ with $f_\alpha^-(x)$ and $f_\alpha^+(x)$ differentiable real-valued functions with respect to x, uniformly with respect to $\alpha \in [0, 1]$. Then*

$$\left[F'_g(x) \right]_\alpha \quad (3.35)$$

$$= \left[\inf_{\beta \geq \alpha} \min\{ \left(f_\beta^- \right)'(x), \left(f_\beta^+ \right)'(x) \}, \sup_{\beta \geq \alpha} \max\{ \left(f_\beta^- \right)', \left(f_\beta^+ \right)'(x) \} \right] \quad (3.36)$$

whenever F is g-differentiable.

Proof. See [10].

Summary of the derivatives for fuzzy-number-valued functions:

- **The GH-, gH-, and g-derivatives generalize the H-derivative.** An H-differentiable function is always GH-, gH-, and g-differentiable.
- **The gH- and g-derivatives generalize the GH-derivative.** A GH-differentiable function is always gH- and g-differentiable.
- **The g-derivative generalizes the gH-derivative.** A gH-differentiable function is always g-differentiable.

3.1.3 Fundamental Theorem of Calculus

Fundamental Theorems of Calculus provide connections between derivatives and integrals, showing that they are inverses of one another.

Theorem 3.8 ([21]). *Let $F : [a, b] \rightarrow \mathscr{F}_{\mathscr{C}}(\mathbb{R}^n)$ be continuous, then $G(x) = \int_a^x F(s)ds$ is H-differentiable and*

$$G'_H(x) = F(x). \tag{3.37}$$

Theorem 3.9 ([21]). *Let $F : [a, b] \rightarrow \mathscr{F}_{\mathscr{C}}(\mathbb{R}^n)$ be H-differentiability and the H-derivative F'_H be integrable over $[a, b]$. Then*

$$F(x) = F(a) + \int_a^x F'_H(s)ds, \tag{3.38}$$

for each $x \in [a, b]$.

The H-differentiable is equivalent to strongly generalized differentiability (i) in Definition 3.6. For the case (ii) in the same definition, Bede and Gal have proved the following theorem.

Theorem 3.10 ([9]). *Let $F : [a, b] \rightarrow \mathscr{F}_{\mathscr{C}}(\mathbb{R})$ be (ii)-differentiable. Then the derivative F'_G is integrable over $[a, b]$ and*

$$F(x) = F(b) - \int_x^b F'_G(s)ds, \tag{3.39}$$

for each $x \in [a, b]$.

3.2 Fuzzy Calculus for Fuzzy Bunches of Functions

The fuzzy calculus for fuzzy bunches of functions, based on the definitions of derivative and integral via extension of the correspondent classical operators, was recently elaborated in [4, 18, 19]. This theory is reviewed and further developed in the present section.

3.2.1 Integral

The integral operator will be represented by $\int$, i.e.,

$$\int : L^1([a,b];\mathbb{R}^n) \to \mathscr{A}C([a,b];\mathbb{R}^n)$$
$$f \mapsto \int_a^t f$$

(3.40)

$t \in [a,b]$ (see Appendix for definitions of spaces of functions).

Definition 3.10 ([3, 18]). Let $F \in \mathscr{F}(L^1([a,b];\mathbb{R}^n))$. The integral of F is given by $\hat{\int}F$, whose membership function is

$$\mu_{\hat{\int}F}(y) = \begin{cases} \sup_{f \in \int^{-1}y} \mu_F(f), & \text{if } \int^{-1}y \neq \emptyset \\ 0, & \text{if } \int^{-1}y = \emptyset \end{cases},$$

(3.41)

for all $y \in \mathscr{A}C([a,b];\mathbb{R}^n)$. In words, $\hat{\int}$ is the extension of the operator $\int$.

The next theorem is a consequence of Theorem 2.6.

Theorem 3.11. *If $F \in \mathscr{F}(L^1([a,b];\mathbb{R}^n))$,*

$$\left[\hat{\int}F\right]_\alpha = \int[F]_\alpha$$
$$= \left\{ \int f : f \in [F]_\alpha \subset L^1([a,b];\mathbb{R}^n) \right\},$$

(3.42)

for all $\alpha \in [0,1]$.

Proof. Since the integral is a continuous operator, the result follows directly from Theorem 2.6.

We next define a linear structure in $\mathscr{F}(L^1([a,b];\mathbb{R}^n))$. Given two fuzzy bunches of functions F and G and $\lambda \in \mathbb{R}$,

$$\mu_{F+G}(h) = \sup_{f+g=h} \min\{\mu_F(f), \mu_G(g)\},$$

(3.43)

$$\mu_{\lambda F}(f) = \begin{cases} \mu_F(h/\lambda) & \text{if } \lambda \neq 0 \\ \chi_0(f) & \text{if } \lambda = 0 \end{cases}.$$

(3.44)

Since these operations are extensions of addition and multiplication by scalar, which are continuous, Theorem 2.6 assures that given $F, G \in \mathscr{F}_{\mathscr{H}}(L^1([a, b]; \mathbb{R}^n))$ and $\lambda \in \mathbb{R}$,

$$F + G \in \mathscr{F}_{\mathscr{H}}(L^1([a, b]; \mathbb{R}^n)) \quad \text{and} \quad [F + G]_\alpha = [F]_\alpha + [G]_\alpha \qquad (3.45)$$

and

$$\lambda F \in \mathscr{F}_{\mathscr{H}}(L^1([a, b]; \mathbb{R}^n)) \quad \text{and} \quad [\lambda F]_\alpha = \lambda [F]_\alpha \qquad (3.46)$$

for all $\alpha \in [0, 1]$.

Theorem 3.12. *Let* $F, G \in \mathscr{F}_{\mathscr{H}}(L^1([a, b]; \mathbb{R}^n))$, *then*

(i) $\hat{\int}(F + G) = \hat{\int}F + \hat{\int}G$;
(ii) $\hat{\int}\lambda F = \lambda\hat{\int}F$, *for any* $\lambda \in \mathbb{R}$.

Proof. From Theorem 2.6 and the linearity of the integral operator,

$$
\begin{aligned}
[\hat{\int}(F + G)]_\alpha &= \int[F + G]_\alpha \\
&= \int\{h : h = f + g, f \in [F]_\alpha, g \in [G]_\alpha\} \\
&= \{\int(f + g), f \in [F]_\alpha, g \in [G]_\alpha\} \\
&= \{\int f + \int g, f \in [F]_\alpha, g \in [G]_\alpha\} \\
&= \{\int f, f \in [F]_\alpha\} + \{\int g, g \in [G]_\alpha\} \\
&= \int[F]_\alpha + \int[G]_\alpha \\
&= [\hat{\int}F]_\alpha + [\hat{\int}G]_\alpha
\end{aligned}
\qquad (3.47)
$$

and

$$
\begin{aligned}
[\hat{\int}\lambda F]_\alpha &= \int[\lambda F]_\alpha \\
&= \{\int \lambda f : f \in [F]_\alpha\} \\
&= \{\lambda \int f : f \in [F]_\alpha\} \\
&= \lambda\{\int f : f \in [F]_\alpha\} \\
&= \lambda \int[F]_\alpha \\
&= \lambda[\hat{\int}F]_\alpha
\end{aligned}
\qquad (3.48)
$$

for all $\alpha \in [0, 1]$.

Example 3.5. Let A be the symmetrical triangular fuzzy number with support $[-a, a]$, $a > 0$. The fuzzy function $F(\cdot) \in \mathscr{F}(L^1([0, T]; \mathbb{R}))$ such that

$$[F(\cdot)]_\alpha = \{f(\cdot) : f(t) = \gamma t, \gamma \in [A]^\alpha\} \qquad (3.49)$$

where $f(\cdot) : [0, T] \to \mathbb{R}$, for each $\alpha \in [0, 1]$, has attainable sets

$$F(t) = At. \tag{3.50}$$

To determine the integral of F using Definition 3.10, one needs to explicit the membership function of A and F:

$$\mu_A(\gamma) = \begin{cases} \dfrac{\gamma}{a} + 1, & \text{if } -a \le \gamma < 0 \\ -\dfrac{\gamma}{a} + 1, & \text{if } 0 \le \gamma < a \\ 0, & \text{otherwise} \end{cases} \tag{3.51}$$

and

$$\mu_F(f) = \begin{cases} \dfrac{\gamma}{a} + 1, & \text{if } f(t) = \gamma t \text{ with } -a \le \gamma < 0 \\ -\dfrac{\gamma}{a} + 1, & \text{if } f(t) = \gamma t \text{ with } 0 \le \gamma < a \\ 0, & \text{otherwise} \end{cases} \tag{3.52}$$

Formula (3.41) states that $\mu_{\hat{\int} F}(y) \ne 0$ only if there exists f such that $\int f = y$ and $\mu_F(f) \ne 0$. In this example, it happens only if $f(t) = \gamma t$ with $\gamma \in [A]^{\alpha}$, that is, $y = \gamma t^2/2$.

$$
\begin{aligned}
\mu_{\hat{\int} F}(\gamma t^2/2) &= \sup\nolimits_{\int f = \gamma t^2/2} \mu_F(f) \\
&= \sup\nolimits_{\int (\gamma t) = \gamma t^2/2} \mu_F(\gamma t) \\
&= \mu_F(\gamma t) \\
&= \begin{cases} \frac{\gamma}{a} + 1, & \text{if } -a \le \gamma < 0 \\ -\frac{\gamma}{a} + 1, & \text{if } 0 \le \gamma < a \\ 0, & \text{otherwise} \end{cases} \\
&= \mu_A(\gamma).
\end{aligned} \tag{3.53}
$$

Hence

$$\mu_{\hat{\int} F}(f) = \begin{cases} \dfrac{\gamma}{a} + 1, & \text{if } f(t) = \gamma t^2/2 \text{ with } -a \le \gamma < 0 \\ -\dfrac{\gamma}{a} + 1, & \text{if } f(t) = \gamma t^2/2 \text{ with } 0 \le \gamma < a \\ 0, & \text{otherwise} \end{cases} \tag{3.54}$$

or

$$[F(\cdot)]_\alpha = \{f(\cdot) : f(t) = \gamma t^2/2, \gamma \in [A]_\alpha\}. \tag{3.55}$$

For each $\alpha \in [0, 1]$, its attainable sets are

$$F(t) = At^2/2. \tag{3.56}$$

The Aumann integral of (3.50) can be calculated levelwise and we obtain the same attainable sets as obtained with $\hat{f}$:

$$
\begin{aligned}
[\textstyle\int F(t)]_\alpha &= [\textstyle\int f_\alpha^-, \int f_\alpha^-] \\
&= [-at^2/2, at^2/2] \\
&= [A]_\alpha t^2/2.
\end{aligned}
\tag{3.57}
$$

The next section introduces the derivative operator for fuzzy bunches of functions. It is defined for more restricted spaces than the integral since they are extensions of the classical case. Also, different from the integral case, we explore the derivative on different spaces (Example 3.9) due to the fact that it is not a continuous operator (in general). We are more interested, though, in differentiating fuzzy bunches of the space of absolutely continuous functions (see Appendix), since we can differentiate more elements in this space than in the space of differentiable functions. Furthermore, it is used and has been explored in the differential inclusions theory, which, as already mentioned, has important connections with the theory we propose to develop.

3.2.2 Derivative

The derivative operator in the sense of distributions (see [1]) will be represented by D, that is,

$$
\begin{aligned}
D : \mathscr{A}C([a, b]; \mathbb{R}^n) &\to L^1([a, b]; \mathbb{R}^n) \\
f &\mapsto Df
\end{aligned}
\tag{3.58}
$$

Thus, there exists $Df(t)$ a.e., in $[a, b]$.

Definition 3.11. Let $F \in \mathscr{F}(\mathscr{A}C([a, b]; \mathbb{R}^n))$. The derivative of F is given by $\hat{D}F$, whose membership function is

$$
\mu_{\hat{D}F}(y) = \begin{cases} \sup_{f \in D^{-1}y} \mu_F(f), & \text{if } D^{-1}y \neq \emptyset \\ 0, & \text{if } D^{-1}y = \emptyset \end{cases}.
\tag{3.59}
$$

for all $y \in L^1([a, b]; \mathbb{R}^n)$. In words, $\hat{D}$ is the extension of operator D.

Example 3.6. Let $F(\cdot)$ be the same fuzzy bunch as in Example 3.5. We note that $F(\cdot) \in \mathscr{F}(\mathscr{A}C([a, b]; \mathbb{R}))$.

Following the same reasoning as Example 3.5,

$$
\begin{aligned}
\mu_{\hat{D}F}(\gamma) &= \sup_{Df=\gamma} \mu_F(f) \\
&= \sup_{D(\gamma t)=\gamma} \mu_F(\gamma t) \\
&= \mu_F(\gamma t) \\
&= \begin{cases} \frac{\gamma}{a} + 1, & \text{if } -a \leq \gamma < 0 \\ -\frac{\gamma}{a} + 1, & \text{if } 0 \leq \gamma < a \\ 0, & \text{otherwise} \end{cases} \\
&= \mu_A(\gamma).
\end{aligned}
\tag{3.60}
$$

It means that the support of $\hat{D}F(\cdot)$ is composed of constant functions such that, at each instant t, the derivative of $F(\cdot)$ is always the fuzzy number A.

Lemma 3.1. *For D defined as above, the preimage $D^{-1}g$ is a closed nonempty subset in the space of functions $\mathscr{A}C([a,b]; \mathbb{R}^n)$ with respect to the uniform norm for each $g \in L^1([a,b]; \mathbb{R}^n)$.*

Proof. $D^{-1}g$ is a finite dimensional subspace of $\mathscr{A}C([a,b]; \mathbb{R}^n)$ since $D^{-1}g = \{f + k : k \in \mathbb{R}^n\}$ for $f \in \mathscr{A}C([a,b]; \mathbb{R}^n)$ such that $f = \int_a^x g$. Hence $D^{-1}g$ is closed.

Theorem 3.13 ([4]). *Let $F \in \mathscr{F}_{\mathscr{K}}(\mathscr{A}C([a,b]; \mathbb{R}^n))$. Then*

$$
[\hat{D}F]_\alpha = D[F]_\alpha.
\tag{3.61}
$$

Proof. This proof will make use of the result: $[F]_0 \cap D^{-1}(g)$ is compact. It is true since the subset $D^{-1}(g)$ is nonempty and it is closed (from Lemma 3.1). Also, $[F]_0 \cap D^{-1}(g)$ is a closed subset of the compact set $[F]_0$, hence it is compact.

We show inclusion $[\hat{D}(F)]_\alpha \subset D([F]_\alpha)$ considering two cases: $\alpha \in (0,1]$ and later $\alpha = 0$.

(i) For $\alpha \in (0,1]$, let $g \in [\hat{D}(F)]_\alpha$, then

$$
\alpha \leq \hat{D}(F)(g) = \sup_{h \in D^{-1}(g)} F(h) = \sup_{h \in [F]_0 \cap D^{-1}(g)} F(h) = F(f)
$$

for some f, since F is an upper semicontinuous function (that is, the membership of F is usc) and $[F]_0 \cap D^{-1}(g)$ is compact. So, $F(f) \geq \alpha$. That is, $f \in [F]_\alpha \cap D^{-1}(g)$. Hence $g \in D([F]_\alpha)$.

(ii) For $\alpha = 0$,

$$
\cup_{\alpha \in (0,1]}[\hat{D}(F)]_\alpha \subset \cup_{\alpha \in (0,1]} D([F]_\alpha) \subseteq D([F]_0).
$$

Consequently,

$$
[\hat{D}(F)]_0 = \overline{\cup_{\alpha \in (0,1]}[\hat{D}(F)]_\alpha} \subset \overline{\cup_{\alpha \in (0,1]} D([F]_\alpha)} \subseteq \overline{D([F]_0)} = D([F]_0).
$$

The last equality holds because D is a closed operator.

Now we prove the inclusion $D([F]_\alpha) \subset [\hat{D}(F)]_\alpha$. If $g \in D([F]_\alpha)$, there exists $f \in [F]_\alpha$ such that $D(f) = g$. Thus,

$$\hat{D}(F)(g) = \sup_{h \in D^{-1}(g)} F(h) \geq F(f) \geq \alpha \Rightarrow g \in [\hat{D}(F)]_\alpha$$

for all $\alpha \in [0, 1]$.

We have proved that $[\hat{D}(F)]_\alpha \subset D([F]_\alpha)$ and $D([F]_\alpha) \subset [\hat{D}(F)]_\alpha$, for all $\alpha \in [0, 1]$, then (3.61) holds.

Example 3.7. Consider $g : [a, b] \to \mathbb{R}$ a differentiable and positive function, $A = (c; d; e)$ a triangular fuzzy number and the fuzzy-number-valued function

$$F(x) = Ag(x). \tag{3.62}$$

We have

$$[F(x)]_\alpha = [f_\alpha^-(x), f_\alpha^+(x)] \tag{3.63}$$

with

$$f_\alpha^-(x) = [a + \alpha(b - a)]g(x) \quad \text{and} \quad f_\alpha^+(x) = [e - \alpha(e - d)]g(x) \tag{3.64}$$

differentiable with respect to x and continuous with respect to α.

The continuity in α means that $F(x) \in \mathscr{F}_C^0(\mathbb{R})$. It will be proved in Theorem 3.17 that the representative bunch of first kind of this function has compact α-cuts in $\mathscr{A}C([a, b]; \mathbb{R})$, since it satisfies the hypotheses of the theorem.

The derivative of the representative bunch of first kind has α-cuts

$$\begin{aligned}
[\hat{D}\tilde{F}]_\alpha &= \bigcup_{\beta \geq \alpha} \bigcup_{0 \leq \lambda \leq 1} (f_\beta^\lambda)' \\
&= (1 - \lambda)[a + \alpha(b - a)]g' + \lambda[e - \alpha(e - d)]g' \\
&= \{a \cdot g', a \in [A]_\alpha\}
\end{aligned} \tag{3.65}$$

for all $\alpha \in [0, 1]$, that is,

$$\hat{D}\tilde{F} = Ag'. \tag{3.66}$$

It is a similar result as in Example 3.2 for GH-derivative, in terms of attainable sets.

Example 3.8. Let

$$f(x) = Be^{cx} \tag{3.67}$$

be a fuzzy-set-valued function where c is a real constant and B is a fuzzy subset in $\mathbb{R}$ such that $B(1) = 1$, $B(0.5) = 0.5$ and $B(x) = 0$ everywhere else. Hence $f(x)$ is not differentiable using Hukuhara or any generalized derivatives since it is not a fuzzy-number-valued function. On the other hand, the fuzzy bunch of functions with α-levels

$$[\tilde{f}(\cdot)]^\alpha = \begin{cases} \{y_1(\cdot), y_2(\cdot)\}, & \text{if } 0 \le \alpha \le 0.5 \\ \{y_1(\cdot)\}, & \text{if } 0.5 < \alpha \le 1 \end{cases},$$

where $y_1(x) = e^{cx}$ and $y_2(x) = 0.5e^{cx}$, has (3.67) as attainable fuzzy sets and is $\hat{D}$-differentiable. Since this α-levels are compact subsets of $\mathscr{A}C([a, b]; \mathbb{R})$, we apply Theorem 3.13 and obtain

$$[\hat{D}f(\cdot)]^\alpha = \begin{cases} \{z_1(\cdot), z_2(\cdot)\}, & \text{if } 0 \le \alpha \le 0.5 \\ \{z_1(\cdot)\}, & \text{if } 0.5 < \alpha \le 1 \end{cases},$$

where $z_1(x) = ce^{cx}$ and $z_2(x) = 0.5ce^{cx}$. Its attainable sets are

$$\hat{D}f(x) = cBe^{cx}. \tag{3.68}$$

Remark 3.2. The Hukuhara or the generalized derivatives cannot be used to differentiate fuzzy-set-valued functions whose images are not fuzzy numbers, as the function in Example 3.8. On the other hand, one can use the $\hat{D}$ on correspondent fuzzy bunches of functions and regard its attainable fuzzy sets as derivative.

Example 3.9 ([4]). The operator $\hat{D} : \mathscr{F}_\mathscr{K}(C^1([a, b]; \mathbb{R}^n)) \to \mathscr{F}_\mathscr{K}(C([a, b]; \mathbb{R}^n))$ is well defined and for each $F \in \mathscr{F}_\mathscr{K}(C^1([a, b]; \mathbb{R}^n))$ we have

$$[\hat{D}F]^\alpha = D[F]^\alpha \tag{3.69}$$

for all $\alpha \in [0, 1]$, if $C^1([a, b]; \mathbb{R}^n)$ is endowed with the norm $\| x \|_1 = \sup_{0 \le t \le T}\{|x(t)| + |x'(t)|\}$ and $C([a, b]; \mathbb{R}^n)$ is endowed with the usual supremum norm. The result follows from Theorem 2.6 since D is a continuous function for these spaces.

Another possibility of D being a continuous operator is as follows:

Theorem 3.14 ([4]). *Consider the subset in $\mathscr{A}C([0, T]; \mathbb{R}^n)$:*

$$Z_T(\mathbb{R}^n) = \{x(\cdot) \in C([0, T]; \mathbb{R}^n) : \exists\, x'(\cdot) \in L^\infty([0, T]; \mathbb{R}^n)\}, \tag{3.70}$$

with $Z_T(\mathbb{R}^n)$ having the uniform norm topology and $L^\infty([0, T]; \mathbb{R}^n)$ with the weak-topology. Thus,*

$$\hat{D} : \mathscr{F}_\mathscr{K}(Z_T(\mathbb{R}^n)) \to \mathscr{F}_\mathscr{K}(L^\infty([0, T]; \mathbb{R}^n)), \tag{3.71}$$

where $\hat{D}$ is the extension of the derivative D, is well defined, that is, for each $F \in \mathscr{F}_{\mathscr{K}}(Z_T(\mathbb{R}^n))$, the α-level $[\hat{D}F]^\alpha$ is a compact subset in $L^\infty([0, T]; \mathbb{R}^n)$ and $[\hat{D}F]^\alpha = D[F]^\alpha$.

Proof. The result follows from the Theorem 2.6 because

$$D : Z_T(\mathbb{R}^n) \to L^\infty([0, T]; \mathbb{R}^n) \tag{3.72}$$

is a continuous linear operator (see [1, p. 104]).

Theorem 3.15. *Let $F, G \in \mathscr{F}_{\mathscr{K}}(\mathscr{A}C([a, b]; \mathbb{R}^n))$, then*

(i) $\hat{D}(F + G) = \hat{D}F + \hat{D}G$;
(ii) $\hat{D}\lambda F = \lambda \hat{D}F$, for any $\lambda \in \mathbb{R}$.

Proof. This proof is completely analogous to the one of Theorem 3.15, due to the linearity of the derivative operator.

3.2.3 Fundamental Theorem of Calculus

A result connects the concepts of derivative and integral for fuzzy bunches of functions as in the classical case and in the fuzzy-set-valued function case.

Theorem 3.16. *Let $F \in \mathscr{F}_{\mathscr{K}}(L^1([a, b]; \mathbb{R}^n))$. Hence*

$$\hat{D}\left(\hat{\textstyle\int} F\right) = F, \tag{3.73}$$

that is,

$$\left[\hat{D}\left(\hat{\textstyle\int} F\right)\right]^\alpha = [F]^\alpha. \tag{3.74}$$

for all $\alpha \in [0, 1]$.

Proof. Since Theorem 3.11 holds,

$$\begin{aligned} [\hat{\textstyle\int} F]_\alpha &= \textstyle\int [F]_\alpha \\ &= \{\textstyle\int f : f \in [F]_\alpha\} \end{aligned} \tag{3.75}$$

for all $\alpha \in [0, 1]$ and $\hat{\textstyle\int} F \in \mathscr{F}_{\mathscr{K}}(\mathscr{A}C([0, T]; \mathbb{R}^n))$. Then Theorem 3.13 holds and,

$$\begin{aligned} [\hat{D}\hat{\textstyle\int} F]_\alpha &= D[\hat{\textstyle\int} F]_\alpha \\ &= \{D \textstyle\int f : f \in [F]_\alpha\} \\ &= [F]_\alpha \end{aligned} \tag{3.76}$$

for all $\alpha \in [0, 1]$.

3.3   Comparison

Different fuzzy bunches of functions may present the same attainable fuzzy sets, that is, more than one fuzzy bunch of functions may correspond to one single fuzzy-set-valued function. Choosing the suitable fuzzy bunch may lead to equivalence of $\hat{D}$ with derivatives for fuzzy-set-valued functions and equivalence of $\hat{\int}$ with integrals for fuzzy-set-valued functions (in terms of attainable sets). This section discloses similarities of the proposed theory with other approaches.

The motivation for this comparison and the definition of the two different fuzzy bunches of functions of Definition 2.16 is what happens to the fuzzy-number-valued functions of Examples 3.3 and 3.4. In the former the gH-derivative does not exist whereas the g-derivative does and in the latter both do not exist. We calculate the $\hat{D}$-derivative of the corresponding fuzzy bunches of the fuzzy-valued functions in Examples 3.3 and 3.4 next. The fuzzy-number-valued functions do not meet the conditions of the theorems to be stated, revealing the importance of the hypotheses of these theorems.

Example 3.10. Recall Examples 2.13 and 3.3 where the representative bunch of first kind are given by the α-cuts

$$[\tilde{F}_1(\cdot)]_\alpha = \begin{cases} \left(\displaystyle\bigcup_{\beta \geq 0.5} \bigcup_{0 \leq \lambda \leq 1} f_\beta^\lambda \right) \bigcup \left(\displaystyle\bigcup_{\alpha \leq \beta \leq 0.5} \bigcup_{0 \leq \lambda \leq 1} g_\beta^\lambda \right), \text{if } 0 \leq \alpha \leq 0.5 \\ \displaystyle\bigcup_{\beta \geq \alpha} \bigcup_{0 \leq \lambda \leq 1} f_\beta^\lambda, \text{if } 0.5 < \alpha \leq 1 \end{cases} \tag{3.77}$$

where

$$\begin{cases} f_\beta^\lambda(\cdot) : f_\beta^\lambda(x) = (1-\lambda)(x^2 - 3 + \beta) + \lambda((2\beta-1)x^2 - 6\beta + 4), \\ g_\beta^\lambda(\cdot) : g_\beta^\lambda(x) = (1-\lambda)(x^2 - 3 + \beta) + \lambda((1-2\beta)x^2 - 2\beta + 2), \end{cases} \tag{3.78}$$

for all $\lambda \in [0,1]$. Since

$$\begin{cases} (f_\beta^\lambda)'(\cdot) : f_\beta^\lambda(x) = (1 - 2\lambda + 2\beta\lambda)2x, \\ (g_\beta^\lambda)'(\cdot) : g_\beta^\lambda(x) = (1 - 2\beta\lambda)2x, \end{cases} \tag{3.79}$$

using Theorem 3.13 to calculate $[\hat{D}\tilde{F}_1(\cdot)]_\alpha$ we obtain

$$\begin{cases} \left(\displaystyle\bigcup_{\beta \geq 0.5} \bigcup_{0 \leq \lambda \leq 1} (f_\beta^\lambda)' \right) \bigcup \left(\displaystyle\bigcup_{\alpha \leq \beta \leq 0.5} \bigcup_{0 \leq \lambda \leq 1} (g_\beta^\lambda)' \right), \text{if } 0 \leq \alpha \leq 0.5 \\ \displaystyle\bigcup_{\beta \geq \alpha} \bigcup_{0 \leq \lambda \leq 1} (f_\beta^\lambda)', \text{if } 0.5 < \alpha \leq 1. \end{cases} \tag{3.80}$$

At $x \in [0, 0.8]$

$$[\hat{D}\tilde{F}_1(x)]_\alpha = [m, M] \tag{3.81}$$

with m as

$$\min\left\{\left(\bigcup_{\beta \geq 0.5}\bigcup_{0 \leq \lambda \leq 1}(f_\beta^\lambda)'(x)\right)\bigcup\left(\bigcup_{\alpha \leq \beta \leq 0.5}\bigcup_{0 \leq \lambda \leq 1}(g_\beta^\lambda)'(x)\right)\right\} = 0 \tag{3.82}$$

if $0 \leq \alpha \leq 0.5$ and

$$\min\left\{\bigcup_{\beta \geq \alpha}\bigcup_{0 \leq \lambda \leq 1}(f_\beta^\lambda)'(x)\right\} = (2\alpha - 1)2x \tag{3.83}$$

if $0.5 < \alpha \leq 1$. And M equals

$$\max\left\{\left(\bigcup_{\beta \geq 0.5}\bigcup_{0 \leq \lambda \leq 1}(f_\beta^\lambda)'(x)\right)\bigcup\left(\bigcup_{\alpha \leq \beta \leq 0.5}\bigcup_{0 \leq \lambda \leq 1}(g_\beta^\lambda)'(x)\right)\right\} = 2x \tag{3.84}$$

if $0 \leq \alpha \leq 0.5$ and

$$\max\left\{\bigcup_{\beta \geq \alpha}\bigcup_{0 \leq \lambda \leq 1}(f_\beta^\lambda)'(x)\right\} = 2x \tag{3.85}$$

if $0.5 < \alpha \leq 1$.

Hence the attainable sets of the $\hat{D}$-derivative are

$$[\hat{D}\tilde{F}_1(x)]_\alpha = \begin{cases} [0, 2x], & \text{if } 0 \leq \alpha \leq 0.5 \\ [(2\alpha - 1)2x, 2x], & \text{if } 0.5 < \alpha \leq 1 \end{cases} \tag{3.86}$$

that is, the same as the g-derivative of the fuzzy-number-valued function F.

Example 3.11. Recall Examples 2.12 and 3.4 where the representative bunch of first kind is given by the α-cuts

$$[F(x)]_\alpha = \begin{cases} [10x^2 - 12, 10x^2 + 2], & \text{if } 0 \leq \alpha \leq 0.5 \\ [-1, 1], & \text{if } 0.5 < \alpha \leq 1 \end{cases} . \tag{3.87}$$

and the representative bunch of second kind is defined by

$$[\tilde{F}_1(\cdot)]_\alpha = \begin{cases} \displaystyle\bigcup_{i=1}^{2}\bigcup_{0\leq\lambda\leq1} y_i^\lambda(\cdot), \text{ if } 0 \leq \alpha \leq 0.5 \\ \displaystyle\bigcup_{0\leq\lambda\leq1} y_1^\lambda(\cdot), \text{ if } 0.5 < \alpha \leq 1 \end{cases} \tag{3.88}$$

where

$$\begin{cases} y_1^\lambda(\cdot) : y_1^\lambda(x) = (1-\lambda)(10x^2 - 12) + \lambda(10x^2 + 2), \\ y_2^\lambda(\cdot) : y_2^\lambda(x) = (1-\lambda)(-1) + \lambda, \\ y_3^\lambda(\cdot) : y_3^\lambda(x) = (1-\lambda)(-1) + \lambda(10x^2 + 2), \\ y_4^\lambda(\cdot) : y_4^\lambda(x) = (1-\lambda)(10x^2 - 12) + \lambda, \end{cases} \tag{3.89}$$

for all $\lambda \in [0, 1]$.

The derivatives of the representative bunch of first kind is given by the α-cuts

$$[\hat{D}\tilde{F}_1(\cdot)]_\alpha = \begin{cases} \displaystyle\bigcup_{i=1}^{2}\bigcup_{0\leq\lambda\leq1} (y_i^\lambda)'(\cdot), \text{ if } 0 \leq \alpha \leq 0.5 \\ \displaystyle\bigcup_{0\leq\lambda\leq1} y_1^\lambda(\cdot), \text{ if } 0.5 < \alpha \leq 1 \end{cases} \tag{3.90}$$

and the representative bunch of second kind is defined by

$$[\hat{D}\tilde{F}_2(\cdot)]_\alpha = \begin{cases} \displaystyle\bigcup_{i=1}^{4}\bigcup_{0\leq\lambda\leq1} (y_i^\lambda)'(\cdot), \text{ if } 0 \leq \alpha \leq 0.5 \\ \{(y_1)'(\cdot)\} \displaystyle\bigcup_{0\leq\lambda\leq1} y_1^\lambda(\cdot), \text{ if } 0.5 < \alpha \leq 1 \end{cases} \tag{3.91}$$

where

$$\begin{cases} (y_1)'(\cdot) : (y_1)'(x) = 20x, \\ (y_2)'(\cdot) : (y_2)'(x) = 0, \\ (y_3)'(\cdot) : (y_3)'(x) = \lambda 20x, \\ (y_4)'(\cdot) : (y_4)'(x) = (1 - \lambda)20x, \end{cases} \tag{3.92}$$

for all $\lambda \in [0, 1]$.

In terms of attainable sets, the derivative of the representative bunch of first kind has attainable sets

$$[\hat{D}\tilde{F}_1(x)]_\alpha = \begin{cases} \{0\} \bigcup \{20x\}, & \text{if } 0 \le \alpha \le 0.5 \\ \{20x\}, & \text{if } 0.5 < \alpha \le 1 \end{cases}.$$ (3.93)

The derivative of the representative bunch of second kind for $x \in [0,1]$ has attainable sets

$$[\hat{D}\tilde{F}_2(x)]_\alpha = \begin{cases} [0, 20x], & \text{if } 0 \le \alpha \le 0.5 \\ \{20x\}, & \text{if } 0.5 < \alpha \le 1 \end{cases}$$ (3.94)

and for $x \in [-1, 0]$,

$$[\hat{D}\tilde{F}_1(x)]_\alpha = \begin{cases} [20x, 0], & \text{if } 0 \le \alpha \le 0.5 \\ \{20x\}, & \text{if } 0.5 < \alpha \le 1 \end{cases}.$$ (3.95)

Hence the derivative of the representative bunch of first kind at each $x \in [-1, 1]$ does not define fuzzy numbers while the derivative of the representative bunch of second kind does.

Example 3.10 illustrates that the $\hat{D}$-derivative of the fuzzy bunch of first kind of the given fuzzy-number-valued function F exists but its attainable sets are not fuzzy numbers (while the gH-derivative of the fuzzy-number-valued function does not exist). The result that we state next regards the necessary conditions for equivalence between the gH-derivative of a fuzzy-number-valued function and the $\hat{D}$-derivative of the corresponding fuzzy bunch of first kind. The result we state later is connected with Example 3.11, that is, it is necessary that the g-derivative exist for the equivalence with the derivative of the representative bunch of second kind. The $\hat{D}$ derivative in this last case provided a fuzzy-number-valued function, which no derivative for fuzzy-number-valued functions that we presented can do.

Theorem 3.17. *Let $F : [a, b] \to \mathscr{F}_{\mathscr{C}}^0(\mathbb{R})$ be such that the functions $f_\alpha^-(x)$ and $f_\alpha^+(x)$ are real-valued functions, differentiable with respect to x, uniformly in $\alpha \in [0, 1]$. Suppose also that one of the following two cases holds:*

(a) $\left(f_\alpha^-\right)'(x)$ is increasing, $\left(f_\alpha^+\right)'(x)$ is decreasing as functions of α, and

$$\left(f_1^-\right)'(x) \le \left(f_1^+\right)'(x),$$ (3.96)

or
(b) $\left(f_\alpha^-\right)'(x)$ is decreasing, $\left(f_\alpha^+\right)'(x)$ is increasing as functions of α, and

$$\left(f_1^+\right)'(x) \le \left(f_1^-\right)'(x).$$ (3.97)

Then F generates a representative bunch of first kind $\tilde{F}(\cdot)$ with compact α-levels and whose $\hat{D}$-derivative has attainable sets

$$\left[\hat{D}\tilde{F}(x)\right]_\alpha = [\min\{(f_\alpha^-)'(x), (f_\alpha^+)'(x)\}, \max\{(f_\alpha^-)'(x), (f_\alpha^+)'(x)\}]. \tag{3.98}$$

In words, the $\hat{D}$-derivative coincides with the gH-derivative at each x.

Proof. We prove that the sets A_α in Definition 2.16 are α-cuts of a fuzzy set in $\mathscr{A}C([a,b];\mathbb{R})$ using the same arguments as in Example 2.11. The only difference is to demonstrate compactness, which we do next. Note that any sequence $(f_{\alpha_i}^{\lambda_i})$ in $\bigcup_{\beta \geq \alpha} \bigcup_{0 \leq \lambda \leq 1} f_\beta^\lambda(\cdot)$ has a convergent subsequence whose limit belongs to $\bigcup_{\beta \geq \alpha} \bigcup_{0 \leq \lambda \leq 1} f_\beta^\lambda(\cdot)$, due to the continuity of $f_\beta^\lambda(\cdot)$ as function of the real parameters λ and β defined on closed intervals (compact subsets) $[0,1]$ and $[\alpha,1]$, respectively. And since $f_\beta^\pm$ are differentiable, so are f_β^λ. According to [14], the differentiability with respect to x, uniformly in $\alpha \in [0,1]$, assures that if a sequence of functions converges to a function f, the sequence of its derivatives converges to f'. Since f is differentiable, it belongs to $\mathscr{A}C([a,b];\mathbb{R})$. As a result, $\bigcup_{\beta \geq \alpha} \bigcup_{0 \leq \lambda \leq 1} f_\beta^\lambda$ is compact in $\mathscr{A}C([a,b];\mathbb{R})$ and it is equal to its closure and hence to A_α.

We next make use of Theorem 3.13 since $\tilde{F} \in \mathscr{F}_{\mathscr{K}}(\mathscr{A}C([a,b];\mathbb{R}))$:

$$[\hat{D}\tilde{F}]_\alpha = D[\tilde{F}]_\alpha$$
$$= \bigcup_{\beta \geq \alpha} \bigcup_{0 \leq \lambda \leq 1} (f_\beta^\lambda)' \tag{3.99}$$

for all $\alpha \in [0,1]$. And we observe that for case (a)

$$\bigcup_{0 \leq \lambda \leq 1} (f_\beta^\lambda)'(x) = [(f_\beta^-)'(x), (f_\beta^+)'(x)] \tag{3.100}$$

and

$$(f_\alpha^-)'(x) \leq (f_\beta^-)'(x) \leq (f_1^-)'(x) \leq (f_1^+)'(x) \leq (f_\beta^+)'(x) \leq (f_\alpha^+)'(x) \tag{3.101}$$

for $0 \leq \alpha \leq \beta \leq 1$,

$$[(f_\beta^-)'(x), (f_\beta^+)'(x)] \subseteq [(f_\alpha^-)'(x), (f_\alpha^+)'(x)]. \tag{3.102}$$

Hence

$$[\hat{D}\tilde{F}(x)]_\alpha = \bigcup_{\beta \geq \alpha} [(f_\beta^-)'(x), (f_\beta^+)'(x)]$$
$$= [(f_\alpha^-)'(x), (f_\alpha^+)'(x)] \tag{3.103}$$

for all $\alpha \in [0,1]$.

Similarly, case (b) leads to

$$[\hat{D}\tilde{F}(x)]_\alpha = [(f_\alpha^+)'(x), (f_\alpha^-)'(x)]. \tag{3.104}$$

As a result we obtain the desired expression,

$$\left[\hat{D}\tilde{F}(x)\right]_\alpha = \left[\min\{(f_\alpha^-)'(x), (f_\alpha^+)'(x)\}, \max\{(f_\alpha^-)'(x), (f_\alpha^+)'(x)\}\right], \tag{3.105}$$

for all $\alpha \in [0, 1]$, which the same as stated in Theorem 3.6 for the gH-derivative.

A similar result for connecting $\hat{D}$-derivative and g-derivative is presented in what follows.

Theorem 3.18. *Let $F \in [a, b] \to \mathscr{F}_C^0(\mathbb{R})$ be a function such that $f_\alpha^-(x)$ and $f_\alpha^+(x)$ are differentiable real-valued functions with respect to x, uniformly with respect to $\alpha \in [0, 1]$. Then F generates a representative bunch of second kind $\tilde{F}(\cdot)$ with compact α-levels and whose $\hat{D}$-derivative has attainable sets with levels $[\hat{D}\tilde{F}(x)]_\alpha$ given by*

$$\left[\inf_{\beta \geq \alpha} \min\left\{(f_\beta^-)'(x), (f_\beta^+)'(x)\right\}, \sup_{\beta \geq \alpha} \max\left\{(f_\beta^-)'(x), (f_\beta^+)'(x)\right\}.\right] \tag{3.106}$$

It means that the values of the g-derivative of $F(x)$ and the attainable sets of the $\hat{D}$-derivative of $\tilde{F}(\cdot)$ coincide in every $x \in [a, b]$, whenever the g-derivative exists.

Proof. Using the same argument of the previous proof, it follows that the resultant B_α in Definition 2.16 are compact sets in $\mathscr{A}C([a, b]; \mathbb{R})$ and are the α-cuts of the representative bunch of second kind of F, $\tilde{F}$. We use Theorem 3.13 and obtain

$$\begin{aligned} [\hat{D}\tilde{F}]_\alpha &= D[\tilde{F}]_\alpha \\ &= \bigcup_{\beta,\gamma \geq \alpha} \bigcup_{0 \leq \lambda \leq 1} (f_{\beta,\gamma}^\lambda)' \end{aligned} \tag{3.107}$$

We will prove that $L = \inf_{\beta,\gamma \geq \alpha} \left\{(f_{\beta,\gamma}^\lambda)'(x)\right\}$ is attained, that is, that there exists a triple $(\bar{\lambda}, \bar{\beta}, \bar{\gamma})$ such that $(f_{\bar{\beta},\bar{\gamma}}^{\bar{\lambda}})'(x) = L$ with $\bar{\beta}, \bar{\gamma} \in [\alpha, 1]$, $\bar{\lambda} \in [0, 1]$. From the definition of infimum, $y \geq L$ if $y \in \bigcup_{\beta,\gamma \geq \alpha} \bigcup_{0 \leq \lambda \leq 1} (f_{\beta,\gamma}^\lambda)'(x)$ and there exists a sequence (y_n), $y_n = (f_{\beta_n,\gamma_n}^{\lambda_n})'(x)$ such that

$$(f_{\beta_n,\gamma_n}^{\lambda_n})'(x) \to L, \quad L \leq (f_{\beta_n,\gamma_n}^{\lambda_n})'(x). \tag{3.108}$$

To the sequence (y_n) in $\mathbb{R}$ there corresponds a sequence $(g_n(\cdot))$ of functions such that $g_n(\cdot) = (f_{\beta_n,\gamma_n}^{\lambda_n})'(\cdot)$. This sequence of functions has a convergent subsequence, since the set is sequentially compact (where we use the same result in [14] as previously used). This subsequence of functions defines a subsequence in (y_n), $y_{n_k} = g_{n_k}(x)$. The subsequence (y_{n_k}) also converges to L. The limit of $g_{n_k}(\cdot)$ is attained for some triple $(\bar{\lambda}, \bar{\beta}, \bar{\gamma})$ and its value in x is

$$(f_{\bar{\beta},\bar{\gamma}}^{\bar{\lambda}})'(x) = \lim g_{n_k}(x) = \lim y_{n_k} = L. \tag{3.109}$$

Similarly we prove that the supremum M is also attained. Now we prove that

$$L = \inf_{\beta \geq \alpha} \min \left\{ (f_{\beta}^-)'(x), (f_{\beta}^+)'(x) \right\}. \tag{3.110}$$

For any $(f_{\beta,\gamma}^{\lambda})'(x)$, we have

$$(f_{\beta}^-)' \leq (f_{\beta,\gamma}^{\lambda})'(x) \leq (f_{\gamma}^+)' \quad \text{or} \quad (f_{\gamma}^+)' \leq (f_{\beta,\gamma}^{\lambda})'(x) \leq (f_{\beta}^+)'. \tag{3.111}$$

Hence

$$\inf_{\beta \geq \alpha} \min \left\{ (f_{\beta}^-)'(x), (f_{\beta}^+)'(x) \right\} \leq \inf_{\beta,\gamma \geq \alpha} \left\{ (f_{\beta,\gamma}^{\lambda})'(x) \right\}. \tag{3.112}$$

Since

$$\bigcup_{\beta \geq \alpha} \left\{ (f_{\beta}^-)'(x), (f_{\beta}^+)'(x) \right\} \subset \bigcup_{\beta,\gamma \geq \alpha} \left\{ (f_{\beta,\gamma}^{\lambda})'(x) \right\} \tag{3.113}$$

the equality of the infimum holds.

Hence the value $L = \inf_{\beta \geq \alpha} \min \left\{ (f_{\beta}^-)'(x), (f_{\beta}^+)'(x) \right\}$ is attained by $(f_{\beta}^-)'(x)$ or $(f_{\beta}^+)'(x)$, for some $\beta \geq \alpha$. The same happens to $M = \sup_{\beta \geq \alpha} \max \left\{ (f_{\beta}^-)'(x), (f_{\beta}^+)'(x) \right\}$. As a consequence, there are four possible cases:

(1) $L = (f_{\beta_1}^-)'(x)$ and $M = (f_{\beta_2}^+)'(x)$ and any value between L and M is attained by $(f_{\beta_1,\beta_2}^{\lambda})'(x)$ for some $\lambda \in [0, 1]$;

(2) $L = (f_{\beta_1}^+)'(x)$ and $M = (f_{\beta_2}^-)'(x)$ and any value between L and M is attained by $(f_{\beta_2,\beta_1}^{\lambda})'(x)$ for some $\lambda \in [0, 1]$;

(3) $L = (f_{\beta_1}^-)'(x)$ and $M = (f_{\beta_2}^-)'(x)$ and any value between L and M is attained by $(f_{\beta_1,\beta_1}^{\lambda})'(x)$ or $(f_{\beta_2,\beta_1}^{\lambda})'(x)$ for some $\lambda \in [0, 1]$.

(4) $L = (f_{\beta_1}^+)'(x)$ and $M = (f_{\beta_2}^+)'(x)$ and any value between L and M is attained by $(f_{\beta_1,\beta_1}^{\lambda})'(x)$ or $(f_{\beta_1,\beta_2}^{\lambda})'(x)$ for some $\lambda \in [0, 1]$.

It proves that all values in

$$\left[\inf_{\beta \geq \alpha} \min \left\{ (f_\beta^-)'(x), (f_\beta^+)'(x) \right\}, \sup_{\beta \geq \alpha} \max \left\{ (f_\beta^-)'(x), (f_\beta^+)'(x) \right\} \right] \tag{3.114}$$

are attained.

Then the same expression as in Theorem 3.7 for g-differentiable functions is found and the desired result is proved.

The attainable sets of the $\hat{\int}$-integral of certain bunches of functions also coincide with integrals for fuzzy-set-valued functions, as it will be stated in Theorem 3.19.

Theorem 3.19. *Let $F : [a, b] \to \mathscr{F}_{\mathscr{C}}^0(\mathbb{R})$ be continuous. Then the $\hat{\int}$-integral of the representative bunch of first kind has attainable fuzzy sets*

$$\left[\hat{\int}_a^x \tilde{F} \right]_\alpha = \left[\int_a^x f_\alpha^-, \int_a^x f_\alpha^+ \right] \tag{3.115}$$

for all $\alpha \in [0, 1]$.

In words, the $\hat{\int}$-integral coincides with the integrals for fuzzy-set-valued functions at each x.

Proof. It is not hard to prove the compacity of A_α (Definition 2.16) in $L^1([a, b]; \mathbb{R})$. This is assured by the arguments previously used in proving compacity in $\mathscr{A}C([a, b]; \mathbb{R})$. Following the reasoning of the previous results one demonstrate that A_α are the α-cuts of a fuzzy subset in $L^1([a, b]; \mathbb{R})$.

We observe that $\int_a^x f_\beta^\lambda$ is well defined and that

$$\int_a^x f_\alpha^- \leq \int_a^x f_\beta^\lambda \quad \text{and} \quad \int_a^x f_\beta^\lambda \leq \int_a^x f_\alpha^+ \tag{3.116}$$

for all $\lambda \in [0, 1]$ and $0 \leq \alpha \leq \beta \leq 1$. Hence we obtain, for all $\alpha \in [0, 1]$,

$$[\hat{\int} \tilde{F}]_\alpha = \bigcup_{\beta \geq \alpha} \bigcup_{\lambda \in [0,1]} \int_a^x f_\beta^\lambda \tag{3.117}$$
$$= [\int f_\alpha^-, \int f_\alpha^+]$$

where the last identity holds due to the continuity of $\int_a^x f_\beta^\lambda(x)$ on λ, β, and x.

Thus, we have proved that the attainable sets of the $\hat{\int}$-integral of $\tilde{F}$ have the same expression of the integrals for fuzzy-set-valued functions at each x.

Summary of the comparison of derivatives and integrals:

- **Equivalence between gH- and $\hat{D}$-derivatives.** The gH-derivative of a certain class of fuzzy-number-valued functions coincides with the attainable sets of the $\hat{D}$-derivative (using the representative bunch of first kind).
- **Equivalence between g- and $\hat{D}$-derivatives.** The g-derivative of a certain class of fuzzy-number-valued functions coincides with the attainable sets of the $\hat{D}$-derivative (using the representative bunch of second kind).
- **Equivalence among integrals.** The Aumann, Riemann, and Henstock integrals of a certain class of fuzzy-number-valued functions coincide with the attainable sets of the $\hat{\int}$-integral (using the representative bunch of first kind).

3.4   Summary

This chapter reviewed fuzzy calculus for fuzzy-set-valued functions and presented the new fuzzy calculus using fuzzy bunches of functions. The concepts and results here displayed are essential for the development of the various approaches of FDEs, to be presented in the next chapter. They are summarized next:

- The Hukuhara derivative is defined for a class of fuzzy-set-valued functions and uses the concept of Hukuhara difference. The strongly generalized Hukuhara derivative, weakly generalized Hukuhara derivative, generalized Hukuhara derivative, and the fuzzy generalized derivative generalize the Hukuhara derivative and are defined for wider classes of fuzzy-number-valued functions.
- The Aumann, Riemann, and Henstock integrals are defined for fuzzy-set-valued functions.
- The derivative and the integral via extension of the derivative and integral operators, denoted by $\hat{D}$ and $\hat{\int}$, are defined for fuzzy bunches of functions.
- The $\hat{D}$-derivative of a class of fuzzy bunches of functions coincides with the generalized derivatives in terms of attainable sets.
- The $\hat{\int}$-integral of a class of fuzzy bunches of functions coincides with the integrals for fuzzy-number-valued functions in terms of attainable sets.

References

1. J.P. Aubin, A. Cellina, *Differential Inclusions: Set-Valued Maps and a Viability Theory* (Springer, Berlin/Heidelberg, 1984)

2. R.J. Aumann, Integrals of set-valued functions. J. Math. Anal. Appl. **12**, 1–12 (1965)
3. L.C. Barros, P.A. Tonelli, A.P. Julião, Cálculo diferencial e integral para funções fuzzy via extensão dos operadores de derivação e integração. Technical Report 6 (2010) [in Portuguese]
4. L.C. Barros, L.T. Gomes, P.A. Tonelli, Fuzzy differential equations: an approach via fuzzification of the derivative operator. Fuzzy Sets Syst. **230**, 39–52 (2013)
5. B. Bede, *Mathematics of Fuzzy Sets and Fuzzy Logic* (Springer, Berlin/Heidelberg, 2013)
6. B. Bede, S.G. Gal, Almost periodic fuzzy-number-valued functions. Fuzzy Sets Syst. **147**, 385–403 (2004)
7. B. Bede, S.G. Gal, Quadrature rules for fuzzy-number-valued functions. Fuzzy Sets Syst. **145**, 359–380 (2004)
8. B. Bede, S.G. Gal, Generalizations of the differentiability of fuzzy-number-valued functions with applications to fuzzy differential equations. Fuzzy Sets Syst. **151**, 581–599 (2005)
9. B. Bede, S.G. Gal, Solutions of fuzzy differential equations based on generalized differentiability. Commun. Math. Anal. **9**, 22–41 (2010)
10. B. Bede, L. Stefanini, Generalized differentiability of fuzzy-valued functions. Fuzzy Sets Syst. **230**, 119–141 (2013)
11. Y. Chalco-Cano, H. Román-Flores, M.D. Jiménez-Gamero, Generalized derivative and π-derivative for set-valued functions. Inf. Sci. **181**, 2177–2188 (2011)
12. S.S.L. Chang, L.A. Zadeh, On fuzzy mapping and control. IEEE Trans. Syst. Man Cybern. **2**, 30–34 (1972)
13. D. Dubois, H. Prade, *Fuzzy Sets and Systems: Theory and Applications* (Academic, Orlando, 1980)
14. O. Frink Jr, Differentiation of Sequences. Bull. Am. Math. Soc. **41**, 553–560 (1935)
15. S.G. Gal, Approximation theory in fuzzy setting, in *Handbook of Analytic-Computational Methods in Applied Mathematics*, chapter 13, ed. by G. A. Anastassiou (Chapman & Hall/CRC, Boca Raton, 2000), pp. 617–666
16. R. Goetschel Jr., W. Voxman, Elementary fuzzy calculus. Fuzzy Sets Syst. **18**, 31–43 (1984)
17. L.T. Gomes, L.C. Barros, A note on the generalized difference and the generalized differentiability. Fuzzy Sets Syst. (2015). doi:10.1016/j.fss.2015.02.015
18. L.T. Gomes, L.C. Barros, Fuzzy calculus via extension of the derivative and integral operators and fuzzy differential equations, in *2012 Annual Meeting of the North American Fuzzy Information Processing Society (NAFIPS)* (IEEE, Berkeley, 2012), pp. 1–5
19. L.T. Gomes, L.C. Barros, Fuzzy differential equations with arithmetic and derivative via Zadeh's extension. Mathware Soft Comput. Mag. **20**, 70–75 (2013)
20. M. Hukuhara, Intégration des applications measurables dont la valeur est un compact convexe. Funkc. Ekvacioj **10**, 205–223 (1967)
21. O. Kaleva, Fuzzy differential equations. Fuzzy Sets Syst. 24, 301–317 (1987)
22. M. Puri, D. Ralescu, Differentials of fuzzy functions. J. Math. Anal. Appl. **91**, 552–558 (1983)
23. M. Puri, D. Ralescu, Fuzzy random variables. J. Math. Anal. Appl. **114**, 409–422 (1986)
24. S. Seikkala, On the fuzzy initial value problem. Fuzzy Sets Syst. **24**, 309–330 (1987)
25. L. Stefanini, B. Bede, Generalized Hukuhara differentiability of interval-valued functions and interval differential equations. Nonlinear Anal. Theory Methods Appl. **71**, 1311–1328 (2009)
26. C. Wu, Z. Gong, On Henstock integral of fuzzy-number-valued functions (I). Fuzzy Sets Syst. **120**, 523–532 (2001)

Chapter 4
Fuzzy Differential Equations

We review some approaches of FDEs in this chapter and propose and explore a new theory of FDEs using the $\hat{D}$-derivative introduced in Sect. 3.2.2. Two theorems of existence of solutions to fuzzy initial value problems (FIVPs) are proved and we compare the theory with other approaches, exemplifying with biological models.

4.1 Approaches of FIVPs

FDEs have been extensively studied after [22] first used this expression in 1980. However, the treated problems were not FDEs, strictly, since they did not explicitly use fuzzy sets. Only after the definition of Hukuhara derivative in 1983 did [19] develop a theory for FDEs proposing an existence and uniqueness theorem for solutions to FIVPs. Simultaneously, [28] built up a similar theory for fuzzy-number-valued functions. The proposal of this chapter is to find x satisfying

$$\begin{cases} X'(t) = F(t, X(t)) \\ X(0) = X_0 \end{cases} \tag{4.1}$$

where F is a function that indicates the rate of change of the state variable X at a given instant t.

The function F is real-valued in the IVP as well as the initial condition and the solution. This can be interpreted as a crisp alternative (unique, given some conditions) for the direction to the state variable to follow, at each instant t.

The fuzzy function F indicates a fuzzy direction to be followed in the FIVP. In this case, there are two interpretations. One can fill this trajectory with different crisp solutions, attaching a membership degree to each of them, or one can fill this trajectory with a function that assigns to each instant t a fuzzy subset (that is, the state variable is fuzzy). In the first approach the solution is a fuzzy bunch

© The Author(s) 2015
L.T. Gomes et al., *Fuzzy Differential Equations in Various Approaches*,
SpringerBriefs in Mathematics, DOI 10.1007/978-3-319-22575-3_4

of functions while in the second one obtains a fuzzy-set-valued function. The use of fuzzy functions as solution to FIVPs is justified by the condition of continuity over the function. Having a fuzzy initial value, a continuous fuzzy function does not change abruptly to nonfuzzy states. On the other hand, if the initial value is not fuzzy, only its parameters, then the FDE means that the solution has fuzzy derivative and it is possible only if it is a fuzzy function.

The theories of FDEs in [19, 28] were developed for fuzzy-set-valued functions. The existence theorem for these fuzzy-valued functions is found in Sect. 4.3 is for this kind of functions. The theory for FDEs using fuzzy bunches is what is original research and is presented in Sect. 4.6.

Other approaches that are not strictly FDEs are the extension of the solution and FDIs. The latter makes use of fuzzy bunches of functions and both are based on solving differential equations. Though they do not explicitly use equality of fuzzy sets, they are explored in this chapter since they have similarities with the approach in Sect. 4.6.

It is important to make it clear what "solution" means in each method. Solutions of different approaches may lie in distinct spaces of fuzzy functions, which makes them incomparable, a priori. In this case, we will compare the respective attainable sets. In what follows it will briefly be explained in order to make the comparisons clearer and will be further explored in the next sections. The various approaches of FIVPs that we treat are displayed in Fig. 4.1, where we stress the use of derivatives by three of them.

4.1.1 Fuzzy Differential Equations with Fuzzy Derivatives

Consider the FIVP

$$\begin{cases} X'(t) = F(t, X(t)) \\ X(0) = X_0 \end{cases}, \tag{4.2}$$

where $F : [0, T] \times \mathscr{F}_{\mathscr{C}}(\mathbb{R}^n) \rightarrow \mathscr{F}_{\mathscr{C}}(\mathbb{R}^n)$ and $X_0 \in \mathscr{F}_{\mathscr{C}}(\mathbb{R}^n)$. A solution, if the derivative is *Hukuhara derivative*, is a continuous fuzzy-set-valued function $X : [0, T] \rightarrow \mathscr{F}_{\mathscr{C}}(\mathbb{R}^n)$ that satisfies $X'(t) = F(t, X(t))$, for all $t \in [0, T]$, and the initial condition $X(0) = X_0$. In case the derivative is the *strongly generalized derivative*, the FIVP is defined only for case $n = 1$ and the solution is a continuous fuzzy-number-valued function $X : [0, T] \rightarrow \mathscr{F}_{\mathscr{C}}(\mathbb{R})$ that satisfies $X'(t) = F(t, X(t))$, for all $t \in [0, T]$, and the initial condition $X(0) = X_0$. If the derivative is $\hat{D}$-*derivative*, the solution is a fuzzy bunch of functions $X(\cdot) \in \mathscr{F}_{\mathscr{H}}(\mathscr{A}C([0, T]; \mathbb{R}^n))$ that satisfies $\hat{D}X(t) = F(t, X(t))$ a.e. in $[0, T]$ and the initial condition. That is, given a solution $X(\cdot)$, its derivative $\hat{D}X(\cdot)$ calculated in t (attainable set in t) must be equal to $F(t, X(t))$, a.e. in $[0, T]$. Moreover, the attainable set at $t = 0$, $X(0)$ must satisfy the initial condition.

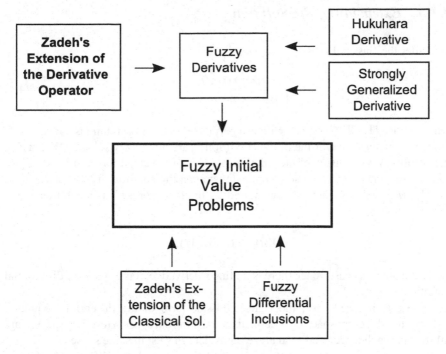

Fig. 4.1 Approaches of FIVPs

4.1.2 *Fuzzy Differential Inclusions*

FDIs are defined levelwise

$$
\begin{aligned}
x'(t) &\in [F(t, x(t))]_\alpha \\
x(0) &\in [X_0]_\alpha
\end{aligned}
\tag{4.3}
$$

for all $\alpha \in [0, 1]$, where $[F]_\alpha : [0, T] \times \mathbb{R}^n \to \mathscr{K}_{\mathscr{C}}^n$ and $[X_0]_\alpha \in \mathscr{K}_{\mathscr{C}}^n$. The solution to (4.3) is a fuzzy bunch of functions $X(\cdot) \in \mathscr{F}_{\mathscr{K}}(\mathscr{A}C([0, T]; \mathbb{R}^n))$ whose elements (functions) of its α-cuts satisfy the differential inclusions (4.3) a.e. in $[0, T]$.

Note that there is no fuzzy derivative. We use the derivative for real-valued functions and there is no equality between fuzzy sets, hence we do not have a *fuzzy* differential equation.

4.1.3 Extension of the Solution

Consider the classical IVP

$$\begin{cases} x'(t) = f(t, x(t), w) \\ x(0) = x_0 \end{cases},$$
(4.4)

where $f : [0, T] \times \mathbb{R}^{n+p} \to \mathbb{R}^n$, with w a parameter in $\mathbb{R}^p$, is continuous and $x_0 \in \mathbb{R}^n$. If the parameter w and/or the initial condition x_0 are now fuzzy subsets (W and X_0), the solution via extension of the solution is a fuzzy-set-valued function $X : [0, T] \to \mathcal{F}(\mathbb{R}^n)$ obtained from the use of the extension on the solution of (4.4) at each $t \in [0, T]$, $x(t, x_0, w)$, that depends on x_0 and w. In other words, X is a solution to the FIVP if

$$X(t) = \hat{x}(t, X_0, W).$$
(4.5)

As in the previous case, since there is no fuzzy derivative, it is not a *fuzzy* differential equation.

Note that in each case, the right-hand-side term of the differential equation belongs to a different space or is defined over a different space. However, to compare all the approaches we need to analyze equivalent FIVPs, in some sense.

First, given the function of the right-hand-side term of the differential equation of one approach, we want to be able to find the corresponding function for the other approaches. Second, we want to compare the five different kinds of FIVPs when they are modeling the same phenomenon. For instance, consider $\lambda \in \mathbb{R}, x \in \mathbb{R}$ and

$$f(x) = \lambda x.$$
(4.6)

It is intuitive to compare IVP (4.4) with this f to FIVP (4.2) having

$$F(X) = \lambda X,$$
(4.7)

$X \in \mathcal{F}(\mathbb{R})$, since the idea of both is that the rate of change of the variable x (or X) is proportional do the variable itself. In this case, the FDI (4.3) is also well defined when $H(x) = \lambda x$.

An option is to consider the extension of f. But, given f, one cannot always find explicitly the expression of F. If $f = x(1 - x)$, the extension of f is not $F(X) = X(1 - X)$, unless the definition of arithmetic via extension principle is not given by Mizumoto and Tanaka [24] and the multiplication carries some kind of interactivity (see Sects. 2.3.2 and 2.3.3). But the Hukuhara and strongly generalized differentiability approaches use other arithmetic to define the derivatives, hence the use of different arithmetics in each side of the equation could be criticized for not being the same. Furthermore, given F, it is not always possible to find a corresponding expression for f [12], as, for example, in the case $[F(X)]_\alpha = [x_0^-, x_\alpha^+]$ for $[X]_\alpha = [x_\alpha^-, x_\alpha^+]$.

In conclusion, "equivalent FIVPs" does not have a clear meaning. In this text we will first take into consideration the interpretation of the FIVP. We will compare the different approaches trying to model the same biological phenomenon.

The solution may belong to different spaces as well. To compare the solutions to the different approaches we do the same as when comparing the derivatives in Sect. 3.3. That is, in case the solution is a fuzzy bunch of functions (via $\hat{D}$-derivative and FDIs), we compare the approaches to the fuzzy attainable sets defined in Sect. 2.5.

This comparison will be carried out for the exponential decay and the logistic models. The classical models are the following IVPs

$$x'(t) = -\lambda x(t), \ x(0) = x_0 \quad \text{(exponential decay)} \tag{4.8}$$

and

$$x'(t) = ax(t)(k - x(t)), \ x(0) = x_0 \quad \text{(logistic equation)} \tag{4.9}$$

with $\lambda, a, k, x_0 > 0$.

We are interested in these models for mathematical and biological reasons. Mathematically, these models are interesting for their equilibrium points: $\bar{x} = 0$ in (4.8) and $\bar{x} = 0$ and $\bar{\bar{x}} = k$ in (4.9). Moreover, $\bar{x} = 0$ in (4.8) and $\bar{\bar{x}} = k$ in (4.9) are asymptotically stable, which means that, given an initial condition in a certain neighborhood, the solution will tend to these equilibrium points. Indeed, any positive initial condition will lead to this behavior. The biological meaning of this is that a population that is represented by (4.8) tends to disappear. This is consistent with the model, since the mathematical equation means that the rate of change is negative and proportional to the existing population. A population which is modeled by (4.9) will tend to the constant k, called carrying capacity (the amount of the population that the environment can support), since if $x < k$ the rate of change is positive and if $x > k$ it is negative.

It is interesting to examine the fuzzy case of these two models because the first one has the extension given by $-\lambda X$, as has already been mentioned. Hence the formula is similar to the classical case. And, as it has been discussed earlier, the extension principle is a good criteria to decide which FIVPs to compare. The expression of the extension of the logistic model, on the other hand, does not have the same representation with the standard arithmetic for fuzzy numbers. But this does not mean that the fuzzy logistic model cannot be written as $X'(t) = AX(t)(K - X(t))$, where all the involved terms are fuzzy. The examples involving the logistic case will use parameters based on the research of Gause (1969), presented as an example by Edelstein-Keshet [15]. Gause carried out an experiment of cultivation of the yeast *Scrhizosaccharomyces kephir* and found out that, beginning with the amount of $x_0 = 0.45$ the population of this yeast clearly satisfied the logistic equation with parameters $k = 5.8$ and $a = 0.01$ for a total time of 160 h.

4.2 Hukuhara Derivative

The authors of [19, 28] independently stated conditions for existence of solution
to differential equations in which the involved functions were fuzzy-set-valued
functions and the derivative was also fuzzy. Both proposed a theorem for existence
and uniqueness of solutions to FIVPs, that is, a Picard–Lindelöf type theorem, one
using the Hukuhara derivative and the other, the Seikkala derivative. The Peano
Theorem was proven not to hold for FDEs because the metric space $(\mathscr{F}_{\mathscr{C}}(\mathbb{R}^n), d_\infty)$
generally is not locally compact [20]. Adding a new condition (boundedness) solved
it [25]. Solution means a fuzzy-set-valued function $X : [0, T] \rightarrow \mathscr{F}_{\mathscr{C}}(\mathbb{R}^n)$ that
satisfies the differential equation for each $t \in [0, T]$ and the initial condition in (1.2).
    Consider (1.2) with the Hukuhara derivative:

$$\begin{cases} X'_H(t) = F(t, X(t)) \\ X(0) = X_0 \end{cases}, \qquad (4.10)$$

where $F : [0, T] \times \mathscr{F}_{\mathscr{C}}(\mathbb{R}) \rightarrow \mathscr{F}_{\mathscr{C}}(\mathbb{R})$ is continuous and $X_0 \in \mathscr{F}_{\mathscr{C}}(\mathbb{R}^n)$. The
following lemma is stated in [19].

Lemma 4.1 ([19]). *A fuzzy-set-valued function $x : [0, T] \rightarrow \mathscr{F}_{\mathscr{C}}(\mathbb{R}^n)$ is a solution
to FIVP (4.10) if and only if it is continuous and satisfies*

$$X(t) = X_0 + \int_0^t F(s, X(s))ds \qquad (4.11)$$

for all $t \in [0, T]$.

    It is not possible to extend Lemma 4.1 for $t < 0$ due to the property of
nondecreasing diameter of the α-levels.

Theorem 4.1 ([25]). *Consider $F : [0, T] \times \mathscr{F}_{\mathscr{C}}(\mathbb{R}^n) \rightarrow \mathscr{F}_{\mathscr{C}}(\mathbb{R}^n)$ continuous and
bounded. Then there is at least one solution to FIVP (1.2) on $[0, T]$.*

    The following result is close to Picard–Lindelöf type theorem, since it establishes
continuity and the Lipschitz condition as sufficient for existence and uniqueness of
solution.

Theorem 4.2 ([19]). *Consider a continuous function $F : [0, T] \times \mathscr{F}_{\mathscr{C}}(\mathbb{R}^n) \rightarrow
\mathscr{F}_{\mathscr{C}}(\mathbb{R}^n)$ satisfying the Lipschitz condition in the second argument, that is, there
exists $k > 0$ such that*

$$d_\infty(F(t, X), F(t, Y)) \leq k d_\infty(X, Y) \qquad (4.12)$$

*for all $t \in [0, T], X, Y \in \mathscr{F}_{\mathscr{C}}(\mathbb{R}^n)$. Then there is a unique solution to FIVP (1.2) on
$[0, T]$.*

A characterization theorem, stated in [6] simplifies the calculations. The result assures that it suffices to solve system of ODEs.

Theorem 4.3 ([6, 7]). *Consider a continuous function* $F : R_0 \to \mathscr{F}_{\mathscr{C}}(\mathbb{R})$, $R_0 = [0, T] \times \overline{B}(X_0, q)$, $q > 0$, $X_0 \in \mathscr{F}_{\mathscr{C}}(\mathbb{R})$, *such that*

$$[F(t,x)]_\alpha = [f_\alpha^-(t,x_\alpha^-,x_\alpha^+), f_\alpha^+(t,x_\alpha^-,x_\alpha^+)], \quad \alpha \in [0,1] \tag{4.13}$$

with $f_\alpha^-(t,x_\alpha^-,x_\alpha^+)$ *and* $f_\alpha^+(t,x_\alpha^-,x_\alpha^+)$ *equicontinuous and uniformly Lipschitz in the second and third arguments, that is, there exists* $L > 0$ *such that*

$$\left| f_\alpha^-(t,x_\alpha^-,x_\alpha^+) - f_\alpha^+(t,x_\alpha^-,x_\alpha^+) \right| \leq L(|x_\alpha^- - y_\alpha^-)| + |x_\alpha^+ - y_\alpha^+)|), \tag{4.14}$$

for any $(t,x), (t,y) \in R_0$ *and for any* $\alpha \in [0,1]$. *Then the FIVP (1.2) has a unique solution in an interval* $[0,k]$, *for some* $k > 0$, *characterized levelwise by the system of ODEs*

$$\begin{cases} (x_\alpha^-)'(t) = f_\alpha^-(t,x_\alpha^-(t),x_\alpha^+(t)) \\ (x_\alpha^+)'(t) = f_\alpha^+(t,x_\alpha^-(t),x_\alpha^+(t)) \\ x_\alpha^-(0) = (x_0^-)_\alpha \\ x_\alpha^+(0) = (x_0^+)_\alpha \end{cases}, \tag{4.15}$$

$\alpha \in [0,1]$.

Some examples with biological interpretation will illustrate the use of Hukuhara derivative in FIVPs.

Example 4.1. Consider the decay model

$$\begin{cases} X_H'(t) = -\lambda X(t) \\ X(0) = X_0 \end{cases}, \tag{4.16}$$

where $\lambda \in \mathbb{R}^+$ and $X_0 \in \mathscr{F}_{\mathscr{C}}(\mathbb{R})$, supp$(X_0) \subset \mathbb{R}^+$. From now on denote $[X_0]_\alpha = [(x_0)_\alpha^-, (x_0)_\alpha^+]$.

The crisp case associated with system (4.16) is frequently used to model population growth (or decay, depending on the sign of λ) or nuclear decay. It is a simple model, yet a reasonable approximation for a short period of observation of the phenomenon. The interpretation of FIVP (4.16) is the decay model with nonfuzzy coefficient and fuzzy initial condition. One explanation is that X_0 can be a label such as "high" or "small" and each real number in $\mathbb{R}$ has a membership degree to this subset. Another interpretation is that there is a partial knowledge of the initial condition, and the most likely values have membership degrees close to one.

Levelwise, solving (4.16) is equivalent to solving

$$\begin{cases} \left[X'_H(t) \right]_\alpha = [-\lambda X(t)]_\alpha \\ [X(0)]_\alpha = [X_0]_\alpha \end{cases} \tag{4.17}$$

for all $\alpha \in [0, 1]$.

Hence,

$$\begin{cases} \left[x_\alpha^-(t), x_\alpha^+(t) \right] = [-\lambda x_\alpha^+(t), -\lambda x_\alpha^-(t)] \\ \left[x_\alpha^-(0), x_\alpha^+(0) \right] = [x_{0\alpha}^-, x_{0\alpha}^+] \end{cases}, \tag{4.18}$$

that is,

$$\begin{cases} (x_\alpha^-(t))' = -\lambda x_\alpha^+(t) \\ (x_\alpha^+(t))' = -\lambda x_\alpha^-(t) \\ x_\alpha^-(0) = x_{0\alpha}^- \\ x_\alpha^+(0) = x_{0\alpha}^+ \end{cases}. \tag{4.19}$$

The solution one obtains is:

$$\begin{cases} x_\alpha^-(t) = c_\alpha^{(1)} e^{\lambda t} + c_\alpha^{(2)} e^{-\lambda t} \\ x_\alpha^+(t) = -c_\alpha^{(1)} e^{\lambda t} + c_\alpha^{(2)} e^{-\lambda t} \end{cases} \tag{4.20}$$

with

$$c_\alpha^- = \frac{x_{0\alpha}^- - x_{0\alpha}^+}{2} \quad \text{and} \quad c_\alpha^+ = \frac{x_{0\alpha}^- + x_{0\alpha}^+}{2}. \tag{4.21}$$

for all $\alpha \in [0, 1]$.

Since it models population or nuclear particles, there is no meaning in the solution when it assumes negative values. This is why it is omitted in Figs. 4.2 and 4.3. The result of nonzero membership degree to negative values is a defect resulting of the Hukuhara derivative. Hence, from $t \approx 40$ on, the solution has no biological meaning anymore, since in the calculations negative values for the state variable are used. Furthermore, in a population that is decreasing proportionally to its quantity it is expected that it tends towards zero, no matter its initial value, uncertainty or its membership to a determined subset ("large," "medium," or "small," for instance). It is expected, actually, that the fuzziness goes to zero. Hence the increasing fuzziness (or diameter) is not considered a good modeling of the decay phenomenon.

Example 4.2. Consider the coefficient of X also fuzzy in the decay model:

$$\begin{cases} X'_H(t) = -\Lambda X(t) \\ X(0) = X_0 \end{cases}, \tag{4.22}$$

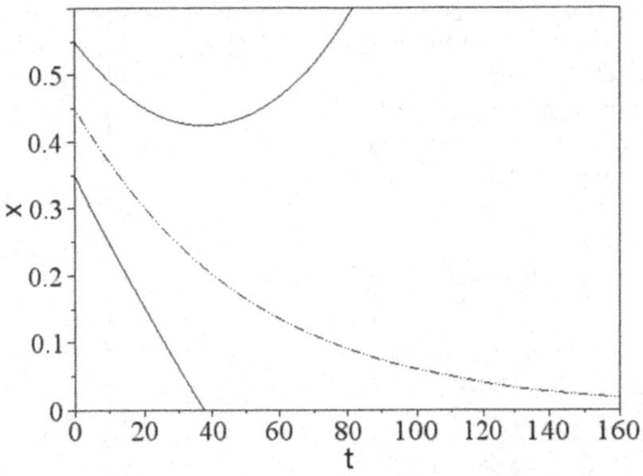

Fig. 4.2 The 0-level (*continuous line*) and the core (*dashed-dotted line*) of solution to the decay model via Hukuhara derivative in Example 4.1. Initial condition (0.35; 0.45; 0.55) and parameter $\lambda = 0.02$

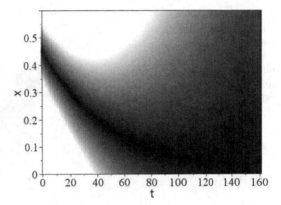

Fig. 4.3 Attainable fuzzy sets of solution to the decay model via Hukuhara derivative in Example 4.1. Initial condition (0.35; 0.45; 0.55) and parameter $\lambda = 0.02$

where $\Lambda \in \mathcal{F}_{\mathscr{C}}(\mathbb{R})$, $\mathrm{supp}(\Lambda) \subset \mathbb{R}^+$, $X_0 \in \mathcal{F}_{\mathscr{C}}(\mathbb{R})$ and $\mathrm{supp}(X_0) \subset \mathbb{R}^+$. From now on assume $[\Lambda]_\alpha = [\lambda_\alpha^-, \lambda_\alpha^+]$.

The biological meaning of this model is the same as the previous one. The sole difference is in the parameter Λ, which is fuzzy in the present example.

Since $[-\Lambda X(t)]_\alpha = [-\lambda_\alpha^+ x_\alpha^+(t), -\lambda_\alpha^- x_\alpha^-(t)]$, the FDE in levels is equivalent to the following system of differential equations

$$\begin{cases} (x_\alpha^-(t))' = -\lambda_\alpha^+ x_\alpha^+(t) \\ (x_\alpha^+(t))' = -\lambda_\alpha^- x_\alpha^-(t) \\ x_\alpha^-(0) = x_{0\alpha}^- \\ x_\alpha^+(0) = x_{0\alpha}^+ \end{cases} \tag{4.23}$$

The solution is

$$\begin{cases} x_\alpha^-(t) = c_\alpha^{(1)} e^{\sqrt{\lambda_\alpha^- \lambda_\alpha^+}\, t} + c_\alpha^{(2)} e^{-\sqrt{\lambda_\alpha^- \lambda_\alpha^+}\, t} \\ x_\alpha^+(t) = -\sqrt{\frac{\lambda_\alpha^-}{\lambda_\alpha^+}} c_\alpha^{(1)} e^{\sqrt{\lambda_\alpha^- \lambda_\alpha^+}\, t} + \sqrt{\frac{\lambda_\alpha^-}{\lambda_\alpha^+}} c_\alpha^{(2)} e^{-\sqrt{\lambda_\alpha^- \lambda_\alpha^+}\, t} \end{cases} \tag{4.24}$$

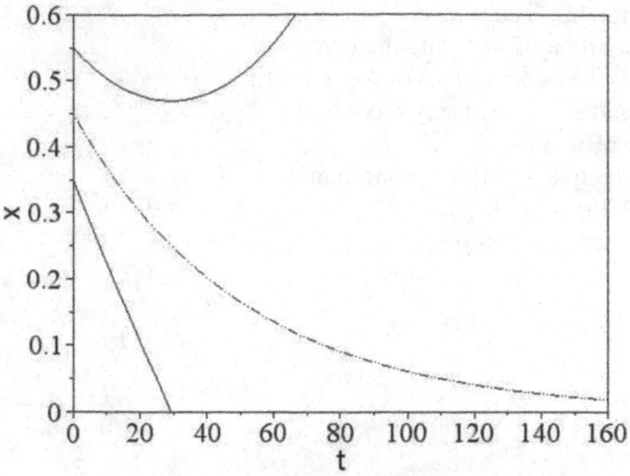

Fig. 4.4 The 0-level
(*continuous line*) and the core
(*dashed-dotted line*) of
solution to the decay model
via Hukuhara derivative in
Example 4.2. Initial condition
(0.35; 0.45; 0.55) and
parameter
$\Lambda = (0.016; 0.020; 0.024)$

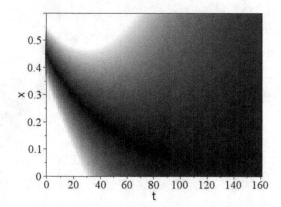

Fig. 4.5 Attainable fuzzy
sets of solution to the decay
model via Hukuhara
derivative in Example 4.2.
Initial condition
(0.35; 0.45; 0.55) and
parameter
$\Lambda = (0.016; 0.020; 0.024)$

with

$$c_\alpha^{(1)} = \frac{x_{0\alpha}^- - \sqrt{\lambda_\alpha^+/\lambda_\alpha^-}\, x_{0\alpha}^+}{2} \quad \text{and} \quad c_\alpha^{(2)} = \frac{x_{0\alpha}^- + \sqrt{\lambda_\alpha^+/\lambda_\alpha^-}\, x_{0\alpha}^+}{2}. \tag{4.25}$$

for all $\alpha \in [0, 1]$.

As in the previous case, no matter the fuzziness of the initial condition or the parameter, it is not expected to increase the diameter of the solution, though it always happens when employing the Hukuhara derivative (Figs. 4.4 and 4.5).

Example 4.3. Another well-known biological model is the logistic growth

$$\begin{cases} X_H'(t) = aX(t)(k - X(t)) \\ X(0) = X_0 \end{cases}. \tag{4.26}$$

Here fuzziness is present only in the initial condition, that is, $X_0 \in \mathscr{F}_{\mathscr{C}}(\mathbb{R})$ and supp$(X_0) \subset \mathbb{R}^+$. The other parameters are nonfuzzy, $a \in \mathbb{R}^+$ and $k \in \mathbb{R}^+$.

The model takes into account that the environment has a limited number of individuals it can support, in a given population. This is characterized by the parameter k, called carrying capacity, here considered constant. Parameter a has to do with reproduction. The term $akX(t)$ corresponds to the growth rate, controlled by the term $-aX(t)^2$, which corresponds to intraspecific competition. If the population is modeled by a crisp variable, note that if $X(t) \approx 0$, $X'_H(t) \approx akX(t)$, that is, there is no obstruction for the population to grow. The change rate is positive while $X(t) < k$, but it tends towards zero while $X(t)$ tends to k. For $X(t) > k$, that is, above the carrying capacity, the change of rate is negative.

The Hukuhara difference is not defined for $k \ominus_H X(t)$ if $X(t)$ is fuzzy. Therefore consider the difference based on SIA (or gH-difference, which gives us the same result for this case). Condition $0 < u < v$ implies

$$\min\{au(k-u), au(k-v), av(k-u), av(k-v)\} = au(k-v) \tag{4.27}$$

and

$$\max\{au(k-u), au(k-v), av(k-u), av(k-v)\} = av(k-u). \tag{4.28}$$

Hence

$$[aX(t)(k - X(t))]_\alpha = [ax_\alpha^-(t)(k - x_\alpha^+(t)), ax_\alpha^+(t)(k - x_\alpha^-(t))] \tag{4.29}$$

since $0 < x_\alpha^-(t) < x_\alpha^+(t)$.

The system of equations

$$\begin{cases} (x_\alpha^-(t))' = ax_\alpha^-(t)(k - x_\alpha^+(t)) \\ (x_\alpha^+(t))' = ax_\alpha^+(t)(k - x_\alpha^-(t)) \\ x_\alpha^-(0) = x_{0\alpha}^- \\ x_\alpha^+(0) = x_{0\alpha}^+ \end{cases} \tag{4.30}$$

is solved numerically, by applying first-order Euler method. This method approximates $x_\alpha^-(t)$, $x_\alpha^+(t)$, $x_\alpha^-(t+h)$ and $x_\alpha^+(t+h)$ by $u_\alpha^{(i)}$, $v_\alpha^{(i)}$, $u_\alpha^{(i+1)}$ and $v_\alpha^{(i+1)}$ such that

$$u_\alpha^{(i+1)} = u_\alpha^{(i)} + h \cdot a u_\alpha^{(i)}(k - v_\alpha^{(i)}) \tag{4.31}$$

and

$$v_\alpha^{(i+1)} = v_\alpha^{(i)} + h \cdot a v_\alpha^{(i)}(k - u_\alpha^{(i)}), \tag{4.32}$$

where $i = 1, 2, \ldots, n$, n is the number of divisions of $[0, T]$ and $h = T/(n-1)$ is the size of each subinterval of $[0, T]$.

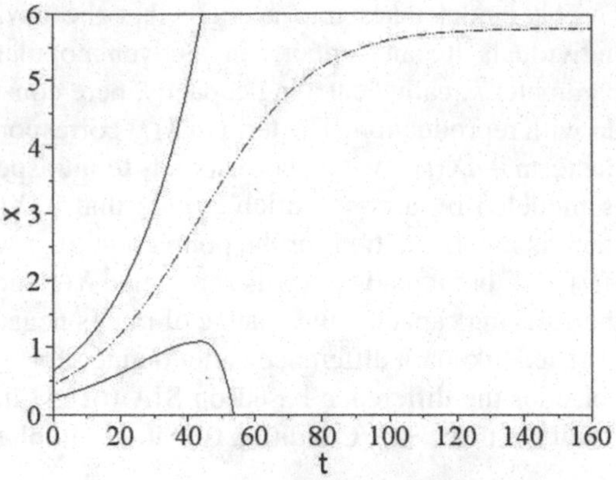

Fig. 4.6 The 0-level (*continuous line*) and the core (*dashed-dotted line*) of solution to the logistic model via Hukuhara derivative in Example 4.3. Initial condition (0.35; 0.45; 0.55) below carrying support $k = 5.8$ and growth parameter $a = 0.01$

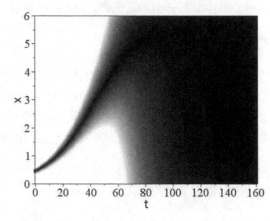

Fig. 4.7 Attainable fuzzy sets of solution to the decay model via Hukuhara derivative in Example 4.3. Initial condition (0.35; 0.45; 0.55) below carrying support $k = 5.8$ and growth parameter $a = 0.01$

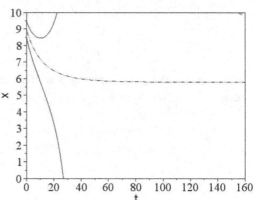

Fig. 4.8 The 0-level (*continuous line*) and the core (*dashed-dotted line*) of solution to the logistic model via Hukuhara derivative in Example 4.3. Initial condition (8.5; 9.0; 9.5) above carrying support $k = 5.8$ and growth parameter $a = 0.01$

The results are illustrated in Figs. 4.6, 4.7, 4.8, and 4.9. As expected, the solution has increasing diameter. Since the core of the initial condition is just one point, the core of the solution is the same as the solution of the crisp case, with initial condition

Fig. 4.9 Attainable fuzzy
sets of solution to the decay
model via Hukuhara
derivative in Example 4.3.
Initial condition
(8.5; 9.0; 9.5) above carrying
support $k = 5.8$ and growth
parameter $a = 0.01$

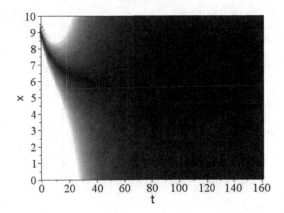

$x_{01}^- = x_{01}^+$. But as in the decay model, it is expected, no matter the initial condition,
from "very small" to "very large," that the population goes to a determined value
(k, in this case) as t increases.

As mentioned at the beginning of this section, other arithmetics can be used.
Though interactivity is not used in the definition of the Hukuhara derivative, let us
see what happens if we admit the CIA to calculate the differential field $aX(k - X)$:

$$[aX(k - X)]_\alpha = [\min_{x \in [X]_\alpha} \{ax(k - x)\}, \max_{x \in [X]_\alpha} \{ax(k - x)\}]. \qquad (4.33)$$

The Euler method is used again, calculating at each step the minimum and the
maximum in the last equation and solving

$$[(x_\alpha^-)'(t), (x_\alpha^+)'(t)] = \left[\min_{x \in [X(t)]_\alpha} \{ax(t)(k - x(t))\}, \max_{x \in [X(t)]_\alpha} \{ax(t)(k - x(t))\} \right].$$
$$\qquad (4.34)$$

Since $f(x) = ax(k - x)$ is a parabola with maximum value at $x = k/2$, it is not
hard to determine the minimum and the maximum at each step:

- If $x_\alpha^-(t) < k/2 < x_\alpha^+(t)$, the maximum of $f(x)$ is attained by $x = k/2$. And
  the minimum is attained at $\min\{ax_\alpha^-(k - x_\alpha^-), ax_\alpha^+(k - x_\alpha^+)\}$.
- If $x_\alpha^+(t) < k/2$, $f(x)$ is increasing with respect to $x \in [x_\alpha^-, x_\alpha^+]$, hence the
  minimum is $f(x_\alpha^-)$ and the maximum is $f(x_\alpha^+)$.
- If $x_\alpha^+(t) < k/2$, $f(x)$ is decreasing with respect to $x \in [x_\alpha^-, x_\alpha^+]$, hence the
  minimum is $f(x_\alpha^+)$ and the maximum is $f(x_\alpha^-)$.

The results are displayed in Fig. 4.10. It is different from the result employing
noninteractive arithmetic, especially when the 0-level starts to assume negative
values. It is mathematically interesting, though biologically it is meaningless from
$t \approx 125$, since negative values are used to calculate the upper 0-level set function
and negative values do not make sense as number of individuals.

Fig. 4.10 The 0-level
(*continuous line*) and the core
(*dashed-dotted line*) of
solution to the logistic model
with CIA and Hukuhara
derivative in Example 4.3.
Initial condition
$(0.35; 0.45; 0.55)$ below
carrying support $k = 5.8$ and
growth parameter $a = 0.01$

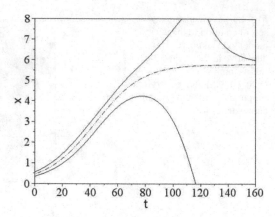

Nevertheless, as mentioned before, it seems incoherent to use different arithmetics to define the derivative and to operate with the right-hand-side function. One would expect thence to define the derivative via CIA.

The FIVPs of the examples in this section satisfy the hypotheses of Theorem 4.3, that is, each α-cut of the function F can be written as function of x_α^- and x_α^+. However, this is not always true and solving the system may become more complicated if this condition is dropped. The reader can refer to [12] for further information about this subject. To briefly illustrate this case, consider the next example.

Example 4.4. Consider FIVP (4.10) with F such that

$$[F(t, X(t))]_\alpha = [x_0^-(t), x_\alpha^+(t)] \tag{4.35}$$

which is equivalent to

$$\begin{cases} (x_\alpha^-)'(t) = x_\alpha^-(t), \ x_\alpha^-(0) = (x_0^-)_\alpha \\ (x_\alpha^+)'(t) = x_\alpha^+(t), \ x_\alpha^+(0) = (x_0^+)_\alpha \end{cases} \tag{4.36}$$

The second equation can be solved directly: $x_\alpha^+(t) = (x_0^+)_\alpha e^t$. The first one needs two steps and we begin with $\alpha = 0$:

$$(x_0^-)'(t) = x_0^-(t), \ x_0^-(0) = (x_0^-)_0 \tag{4.37}$$

which leads to $x_0^-(t) = (x_0^-)_0 e^t$ and

$$(x_\alpha^-)'(t) = (x_0^-)_0 e^t, \ x_\alpha^-(0) = (x_0^-)_\alpha \tag{4.38}$$

The result is

$$\begin{cases} x_\alpha^-(t) = (x_0^-)_0 e^t + (x_0^-)_\alpha - (x_0^-)_0 \\ x_\alpha^+(t) = (x_0^+)_\alpha e^t \end{cases}. \tag{4.39}$$

Example 4.4 illustrated the fact that the function F in FIVP (4.10) may not be written directly as function of x_α^- and x_α^+ (in this example, F is function of x_0^- and x_α^+). The process of solving may become more difficult but since F satisfies the existence Theorem 4.2, there is a solution (and we managed to find it).

4.3 Strongly Generalized Derivative

Section 3.1.2 has shown that the strongly generalized derivative (see Definition 3.6) "fixes" the defect of nondecreasing length of the support of a H-differentiable fuzzy function. In this section we present the existence and uniqueness of two solution theorem, first stated by Bede and Gal [8]. As in the Hukuhara case, the solution to an FIVP is given by integral equations, in such manner that, provided some conditions, there is always a solution with increasing diameter (strongly generalized differentiability of type (i)) and other with decreasing diameter (strongly generalized differentiability of type (ii)) [8]. The possibility of change of type of differentiability (i)–(iv) characterizes interesting phenomena called *switch points*.

Consider (1.2) with the strongly generalized derivative:

$$\begin{cases} X_G'(t) = F(t, X(t)) \\ X(0) = X_0 \end{cases}, \tag{4.40}$$

where $F : \mathbb{R} \times \mathscr{F}_{\mathscr{C}}(\mathbb{R}) \to \mathscr{F}_{\mathscr{C}}(\mathbb{R})$ is continuous and $X_0 \in \mathscr{F}_{\mathscr{C}}(\mathbb{R})$. As in Hukuhara derivative case, there is a result connecting FDEs with fuzzy integral equations.

Theorem 4.4 ([8]). *The FIVP (4.40) is equivalent to the integral equation*

$$X(t) = X_0 + \int_0^t F(s, X(s))ds, \tag{4.41}$$

if the derivative considered is type (i), or to the integral equation

$$X_0 = X(t) + (-1) \int_0^t F(s, X(s))ds, \tag{4.42}$$

if the derivative considered is type (ii), on some interval $[t_1, t_2] \subset [0, T]$.

Based on the next lemma, Bede [7] proves the existence and uniqueness of two solutions (Theorem 4.5).

Lemma 4.2 ([7]). *Let* $X \in \mathscr{F}_{\mathscr{C}}(\mathbb{R})$ *be such that* $[X]_\alpha = [x_\alpha^-, x_\alpha^+]$, $\alpha \in [0, 1]$, x_α^- *and* x_α^+ *differentiable, with* x^- *strictly increasing on* $[0, 1]$, *such that there exist the constants* $c_1 > 0$, $c_2 < 0$ *satisfying* $(x_\alpha^-)' \geq c_1$ *and* $(x_\alpha^+)' \leq c_2$ *for all* $\alpha \in [0, 1]$.

Let $F : [a, b] \rightarrow \mathscr{F}_{\mathscr{C}}(\mathbb{R})$ *be continuous with respect to t, having the level sets* $[F(t)]_\alpha = [f_\alpha^-(t), f_\alpha^+(t)]$ *with bounded partial derivatives* $\frac{\partial f_\alpha^-(t)}{\partial \alpha}$ *and* $\frac{\partial f_\alpha^+(t)}{\partial \alpha}$, *for all* $t \in [a, b]$.

If one of the following two cases occurs

(a) $x_1^- < x_1^+$ *or*
(b) $x_1^- = x_1^+$ *and the core* $[F(s)]_1$ *consists of exactly one element for any* $s \in [a, b]$,

then there exists $h > a$ *such that the H-difference*

$$X \ominus \int_a^t F(s)ds \qquad (4.43)$$

exists for any $t \in [a, h]$.

Theorem 4.5 ([7]). *Let* $R_0 = [0, T] \times \overline{B}(X_0, q)$, $q > 0$, $X_0 \in \mathscr{F}_{\mathscr{C}}(\mathbb{R})$ *and* $F : R_0 \rightarrow \mathscr{F}_{\mathscr{C}}(\mathbb{R})$ *be continuous such that the following assumptions hold:*

(i) There exists a constant $L > 0$ *such that*

$$d_\infty(F(t, X), F(t, Y)) \leq L d_\infty(X, Y) \qquad (4.44)$$

for all $(t, X), (t, Y) \in R_0$.

(ii) Let $[F(t, X)]_\alpha = [f_\alpha^-(t, X), f_\alpha^+(t, X)]$ *be the level set representation of F, then* $f_\alpha^-, f_\alpha^+ : R_0 \rightarrow \mathbb{R}$ *have bounded partial derivatives with respect to* $\alpha \in [0, 1]$, *the bounds being independent of* $(t, X) \in R_0$ *and* $\alpha \in [0, 1]$.

(iii) The functions x_0^- *and* x_0^+ *are differentiable (as functions of* α*), existing* $c_1 > 0$ *with* $(x_0^-)'_\alpha \geq c_1$, *and* $c_2 < 0$ *with* $(x_0^+)'_\alpha \leq c_2$, *for all* $\alpha \in [0, 1]$, *and we have the following possibilities*

(a) $(x_0)_1^- < (x_0)_1^+$
    or
(b) if $(x_0)_1^- = (x_0)_1^+$, *then the core* $[F(t, X)]_1$ *consists in exactly one element for any* $(t, X) \in R_0$, *whenever* $[X]_1$ *consists in exactly one element.*

Then the FIVP (1.2) has exactly two solutions on some interval $[0, k]$, $k > 0$.

The solution of FIVPs using strongly generalized differentiability, as in Hukuhara case, can also be obtained by solving systems of ODEs to find the level set functions of the solution.

Theorem 4.6 ([7]). *Let* $R_0 = [0, T] \times \overline{B}(X_0, q)$, $q > 0$, $X_0 \in \mathscr{F}_{\mathscr{C}}(\mathbb{R})$ *and* $F : R_0 \rightarrow \mathscr{F}_{\mathscr{C}}(\mathbb{R})$ *be such that*

$$[F(t, X)]_\alpha = [f_\alpha^-(t, x_\alpha^-, x_\alpha^+), f_\alpha^+(t, x_\alpha^-, x_\alpha^+)], \ \forall \alpha \in [0, 1] \qquad (4.45)$$

and the following assumptions hold:

(i) $f_\alpha^\pm(t, x_\alpha^-, x_\alpha^+)$ *are equicontinuous, uniformly Lipschitz in their second and third arguments, that is, there exists a constant $L > 0$ such that*

$$|f_\alpha^\pm(t, x_\alpha^-, x_\alpha^+) - f_\alpha^\pm(t, y_\alpha^-, y_\alpha^+)| \leq L(|x_\alpha^- - y_\alpha^-| + |x_\alpha^+ - y_\alpha^+|), \qquad (4.46)$$

$\forall (t, X), (t, Y) \in R_0, \alpha \in [0, 1]$.

(ii) $f_\alpha^-, f_\alpha^+ : R_0 \to \mathbb{R}$ *have bounded partial derivatives with respect to $\alpha \in [0, 1]$, the bounds being independent of $(t, X) \in R_0$ and $\alpha \in [0, 1]$.*

(iii) *The functions x_0^- and x_0^+ are differentiable, existing $c_1 > 0$ with $\left(x_0^-\right)_\alpha' \geq c_1$, and $c_2 < 0$ with $\left(x_0^+\right)_\alpha' \leq c_2$, for all $\alpha \in [0, 1]$, and we have the following possibilities*

(a) $(x_0)_1^- < (x_0)_1^+$

or

(b) *if $(x_0)_1^- = (x_0)_1^+$ then the core $[F(t, X)]_1$ consists in exactly one element for any $(t, X) \in R_0$, whenever $[X]_1$ consists in exactly one element.*

Then the FIVP (1.2) is equivalent on some interval $[t_0, t_0 + k]$ with the union of the following two ODEs:

$$\begin{cases} \left(x_\alpha^-\right)'(t) = f_\alpha^-(t, x_\alpha^-(t), x_\alpha^+(t)) \\ \left(x_\alpha^+\right)'(t) = f_\alpha^+(t, x_\alpha^-(t), x_\alpha^+(t)) \quad , \alpha \in [0, 1] \\ x_\alpha^-(t_0) = (x_0)_\alpha^- , x_\alpha^+(t_0) = (x_0)_\alpha^+ \end{cases} \qquad (4.47)$$

$$\begin{cases} \left(x_\alpha^-\right)'(t) = f_\alpha^+(t, x_\alpha^-(t), x_\alpha^+(t)) \\ \left(x_\alpha^+\right)'(t) = f_\alpha^-(t, x_\alpha^-(t), x_\alpha^+(t)) \quad , \alpha \in [0, 1]. \\ x_\alpha^-(t_0) = (x_0)_\alpha^- , x_\alpha^+(t_0) = (x_0)_\alpha^+ \end{cases} \qquad (4.48)$$

Example 4.5. Consider the decay model

$$\begin{cases} X_G'(t) = -\lambda X(t) \\ X(0) = X_0 \end{cases}, \qquad (4.49)$$

where $\lambda \in \mathbb{R}^+$, $X_0 \in \mathscr{F}_\mathscr{C}(\mathbb{R})$ and $\mathrm{supp}(X_0) \subset \mathbb{R}^+$.

The (i)-differentiable solution is Hukuhara differentiable solution, the same as in Example 4.1. The other solution is the one obtained solving

$$\begin{cases} (x_\alpha^-(t))' = -\lambda x_\alpha^-(t) \\ (x_\alpha^+(t))' = -\lambda x_\alpha^+(t) \\ x_\alpha^-(0) = x_{0\alpha}^- \\ x_\alpha^+(0) = x_{0\alpha}^+ \end{cases}. \qquad (4.50)$$

The solution is

$$
\begin{cases}
x_\alpha^-(t) = x_{0\alpha}^- e^{-\lambda t} \\
x_\alpha^+(t) = x_{0\alpha}^+ e^{-\lambda t}
\end{cases}
\tag{4.51}
$$

for all $\alpha \in [0, 1]$.

The (ii)-differentiable solution is a good alternative to the Hukuhara differentiable solution, since it "fixes" the defect of increasing diameter of the solution of the decay model. It is biologically more meaningful in Example 4.5 since it is expected that the uncertainty vanishes as the population gets close to zero (Figs. 4.11 and 4.12).

Example 4.6. As in Example 4.2, consider the coefficient of X also fuzzy in the decay model:

Fig. 4.11 Solutions to the decay model via strongly generalized derivative in Example 4.5: the 0-level (*continuous line*) of the (i)-differentiable solution (in the strongly generalized sense), the 0-level (*dashed line*) of the (ii)-differentiable solution and the core (*dashed-dotted line*) of both. Initial condition (0.35; 0.45; 0.55) and parameter $\lambda = 0.02$

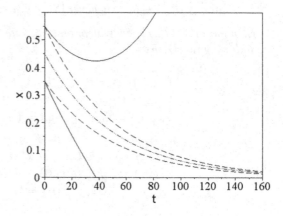

Fig. 4.12 Attainable fuzzy sets of the (ii)-differentiable solution to the decay model via strongly generalized derivative in Example 4.5. Initial condition (0.35; 0.45; 0.55) and parameter $\lambda = 0.02$

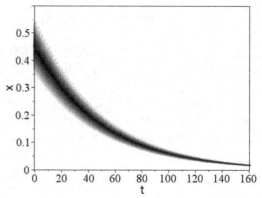

$$\begin{cases} X'_G(t) = -\Lambda X(t) \\ X(0) = X_0 \end{cases} ,$$
(4.52)

where $\Lambda \in \mathscr{F}_{\mathscr{C}}(\mathbb{R})$, supp$(\Lambda) \subset \mathbb{R}^+$, $X_0 \in \mathscr{F}_{\mathscr{C}}(\mathbb{R})$ and supp$(X_0) \subset \mathbb{R}^+$.

The (i)-differentiable solution is Hukuhara differentiable solution, the same as in Example 4.2. The other solution is obtained by solving

$$\begin{cases} (x_\alpha^-(t))' = -\lambda_\alpha^- x_\alpha^-(t) \\ (x_\alpha^+(t))' = -\lambda_\alpha^+ x_\alpha^+(t) \\ x_\alpha^-(0) = x_{0\alpha}^- \\ x_\alpha^+(0) = x_{0\alpha}^+ \end{cases} .$$
(4.53)

The solution is

$$\begin{cases} x_\alpha^-(t) = x_{0\alpha}^- e^{-\lambda_\alpha^- t} \\ x_\alpha^+(t) = x_{0\alpha}^+ e^{-\lambda_\alpha^+ t} \end{cases} .$$
(4.54)

This solution is defined while $x_\alpha^-(t) < x_\alpha^+(t)$, that is, for

$$t < T_m = \frac{1}{\lambda_\alpha^+ - \lambda_\alpha^-} \ln\left(\frac{x_{0\alpha}^+}{x_{0\alpha}^-}\right).$$
(4.55)

The solution also has to satisfy $x_\alpha^-(t) \le x_\beta^-(t)$ and $x_\alpha^+(t) \ge x_\beta^+(t)$ for $0 \le \alpha \le \beta \le 1$. For the same values as Example 4.2, that is, $X_0 = (0.35; 0.45; 0.55)$ and $\Lambda = (0.016; 0.020; 0.024)$, numerically we find $T_m \approx 48$. The solution is displayed in Figs. 4.13 and 4.14. It is clear that the solution does not exist from $T_m \approx 44$ on.

The condition $x_\alpha^-(t) < x_\alpha^+(t)$ restricts the domain for which the solution is defined. This is a particular problem of the solution with decreasing diameter, that is, the Hukuhara differentiable solution does not degenerate and hence its domain is bigger. On the other hand, since this particular problem is an application in population modeling, it does not make sense (biologically) that $x_\alpha^-(t) < 0$, what limits the Hukuhara solution as well.

Note also that if the initial condition is nonfuzzy with $x_\alpha^-(t) = x_\alpha^+(t)$, then $\ln\left(\frac{x_{0\alpha}^+}{x_{0\alpha}^-}\right) = 0$ and hence there is no domain for the solution. Since the crisp condition violates a hypothesis of Theorem 4.5, it cannot assure existence of a solution.

Example 4.7. The FIVP

$$\begin{cases} X'(t) = aX(t)(k - X(t)) \\ X(0) = X_0 \end{cases} .$$
(4.56)

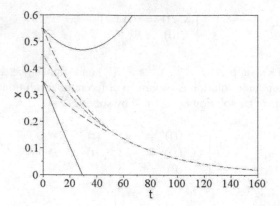

Fig. 4.13 Solutions to the decay model via strongly generalized derivative in Example 4.6: the 0-level (*continuous line*) of the (i)-differentiable solution (in the strongly generalized sense), the 0-level (*dashed line*) of the (ii)-differentiable solution, defined for $t < T_m$, $T_m \approx 48$, and the core (*dashed-dotted line*) of both (for the (ii)-differentiable solution, it is defined for $t < T_m$). Initial condition $(0.35; 0.45; 0.55)$ and parameter $\Lambda = (0.016; 0.020; 0.024)$

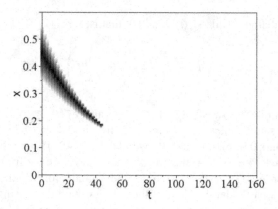

Fig. 4.14 Attainable fuzzy sets of the (ii)-differentiable solution to the decay model via strongly generalized derivative in Example 4.6, which is defined for $t < T_m$, $T_m \approx 48$. Initial condition $(0.35; 0.45; 0.55)$ and parameter $\Lambda = (0.016; 0.020; 0.024)$

where $X_0 \in \mathcal{F}_\mathcal{C}(\mathbb{R})$ and $\text{supp}(X_0) \subset \mathbb{R}^+$, $a \in \mathbb{R}^+$ and $k \in \mathbb{R}^+$ was evaluated numerically in Example 4.3 using the Hukuhara derivative (which is equal to (i)-differentiability).

The (ii)-differentiable solution is obtained considering

$$\begin{cases} (x_\alpha^-(t))' = a x_\alpha^+(t)(k - x_\alpha^-(t)) \\ (x_\alpha^+(t))' = a x_\alpha^-(t)(k - x_\alpha^+(t)) \\ x_\alpha^-(0) = x_{0\alpha}^- \\ x_\alpha^+(0) = x_{0\alpha}^+ \end{cases}. \tag{4.57}$$

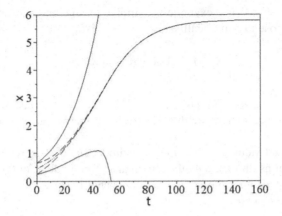

Fig. 4.15 Solutions to the logistic model via strongly generalized derivative in Example 4.7: the 0-level (*continuous line*) of the (i)-differentiable solution (in the strongly generalized sense), the 0-level (*dashed line*) of the (ii)-differentiable solution and the core (*dashed-dotted line*) of both. Initial condition (0.35; 0.45; 0.55) below carrying support $k = 5.8$ and growth parameter $a = 0.01$

In order to solve it numerically, we approximate $x_\alpha^-(t)$, $x_\alpha^+(t)$, $x_\alpha^-(t + h)$ and $x_\alpha^+(t + h)$ by $u_\alpha^{(i)}$, $v_\alpha^{(i)}$, $u_\alpha^{(i+1)}$ and $v_\alpha^{(i+1)}$ such that

$$u_\alpha^{(i+1)} = u_\alpha^{(i)} + h a \, v_\alpha^{(i)}(k - u_\alpha^{(i)}) \tag{4.58}$$

and

$$v_\alpha^{(i+1)} = v_\alpha^{(i)} + h a \, u_\alpha^{(i)}(k - v_\alpha^{(i)}), \tag{4.59}$$

where $i = 1, 2, \ldots, n$, n is the number of divisions of $[0, T]$ and $h = T/(n - 1)$ is the size of each subinterval of $[0, T]$.

The diameter of the solution in the previous example decreases and the function tends to k (see Fig. 4.15). As it was earlier observed, no matter the positive initial condition, the trajectory of the classical case always tends towards k, and that is what is expected from (4.56), considering the phenomenon that originated it. Hence it does not matter the initial fuzziness as well, it is *certain* that, after a certain time, the state variable will be very close to the carrying capacity. With this point of view the (ii)-differentiable solution is more appropriate.

Example 4.8 ([7, 9]). This example is similar to the one presented in [7, 9]. The authors propose to numerically solve

$$\begin{cases} X'_H(t) = AX(t)(K \ominus_{gH} X(t)) \\ X(0) = x_0 \end{cases}, \tag{4.60}$$

where $x_0 \in \mathbb{R}^0$, $K \in \mathscr{F}_\mathscr{C}(\mathbb{R})$, supp$(K) \subset \mathbb{R}^+$, $A \in \mathscr{F}_\mathscr{C}(\mathbb{R})$ and supp$(A) \subset \mathbb{R}^+$. The values of the parameters are different from [7, 9]. We set $[K]_\alpha = [k_\alpha^-, k_\alpha^+]$ and $[A]_\alpha = [a_\alpha^-, a_\alpha^+]$

The crisp initial value does not meet condition (iii) of Theorem 4.5, that guarantees two solutions. In fact, only the Hukuhara differentiable (or (i)-differentiable) solution is admitted, since the initial condition is not fuzzy.

We obtain it by solving

$$\begin{cases} \left(x_\alpha^-\right)'(t) = f_\alpha^-(t, x_\alpha^-(t), x_\alpha^+(t)) \\ \left(x_\alpha^+\right)'(t) = f_\alpha^+(t, x_\alpha^-(t), x_\alpha^+(t)) \quad , \alpha \in [0, 1] \\ x_\alpha^-(t_0) = (x_0)_\alpha^- , x_\alpha^+(t_0) = (x_0)_\alpha^+ \end{cases} \tag{4.61}$$

To obtain the expression of f_α^- and f_α^+, note that the supports of A and X are positive, in order to preserve the biological meaning. Hence

$$[AX]_\alpha = [A]_\alpha [X]_\alpha = [a_\alpha^- x_\alpha^-, a_\alpha^+ x_\alpha^+]. \tag{4.62}$$

Also,

$$[K \ominus_{gH} X]_\alpha = [\min\{k_\alpha^- - x_\alpha^-, k_\alpha^+ - x_\alpha^+\}, \max\{k_\alpha^- - x_\alpha^-, k_\alpha^+ - x_\alpha^+\}], \tag{4.63}$$

provided $K \ominus_{gH} X$ defines a fuzzy number. As a result,

$$\begin{cases} f_\alpha^-(t, x_\alpha^-, x_\alpha^+) = \min_{s,p \in \{-,+\}} \{a_\alpha^s x_\alpha^s (k_\alpha^p - x_\alpha^p)\} \\ f_\alpha^+(t, x_\alpha^-, x_\alpha^+) = \max_{s,p \in \{-,+\}} \{a_\alpha^s x_\alpha^s (k_\alpha^p - x_\alpha^p)\} \end{cases}, \alpha \in [0, 1]. \tag{4.64}$$

We solve it using the Euler method, calculating at each step the minimum and maximum needed to determine f_α^- and f_α^+. If at some point $(x_\alpha^-)'(t) = (x_\alpha^+)'(t)$, that is, $f_\alpha^- = f_\alpha^+$, then the solution would be (iv)-differentiable, according to Definition 3.6. Hence, considering this point as new initial condition, from this point on it would have decreasing diameter or increasing diameter, that is, be (i)-differentiable or (ii)-differentiable. The new problem would meet all conditions of Theorem 4.5 and we would be able to find these two solutions.

This point at which there are two options of solutions, each one with a different differentiability, is called switch point. With the parameters used in this example, there is no such point. Hence, there exists only the Hukuhara differentiable solution (Fig. 4.16).

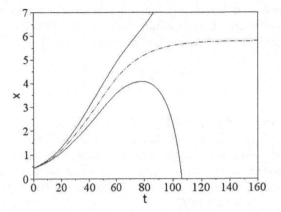

Fig. 4.16 The 0-level (*continuous line*) and the core (*dashed-dotted line*) of the (i)-differentiable solution to the logistic model via strongly generalized derivative in Example 4.8. Initial condition 0.45 below carrying support $K = (5.3; 5.8; 6.3)$ and growth parameter $A = (0.005; 0.010; 0.015)$

4.4 Fuzzy Differential Inclusions

The theory of differential inclusions was developed to deal with some kinds of uncertainties not described by classical dynamical systems. These uncertainties are due, for instance, to partial knowledge arisen from the impossibility of total understanding of a phenomenon or to the ignorance of laws related to the control of the system. Control can be direction, acceleration, fuel, temperature, weight or other variables that may affect the system.

The mathematical model involves a family of differential equations

$$x'(t) = f(x(t), u(t)), \quad u(t) \in U((x(t)), \tag{4.65}$$

where $x \in \mathbb{R}^n$ is the state variable, $u \in \mathbb{R}^m$ is the control, and U is the subset of admissible controls. Together with x, u defines the velocity of the system.

Defining the set-valued map $H : \mathbb{R}^n \to \mathscr{P}(\mathbb{R}^n)$ as $H(x) = f(x, U(x)) = \{f(x, u) : u \in U(x)\}$, the equation in (4.65) can be rewritten as

$$x'(t) \in H(x(t), u(t)). \tag{4.66}$$

Finding a solution x means finding an everywhere differentiable function that satisfies (4.65) and a given initial condition. This problem is said to be *parameterizable*, that is, there exists a single-valued function f of two variables x, u such that, for every x, $H(x) = f(x, U)$. The initial condition can also assume values in a given set in $\mathbb{R}^n$. Being parameterizable is an important property, since f continuous implies that, for every fixed $u^0 \in U$, $x \mapsto f(x, u^0)$ is a continuous selection. The existence of a continuous selection guarantees at least one solution to the Problem (4.66),

that is, an everywhere differentiable function that satisfies the inclusion, obtained
by solving the (4.65) (and given an initial condition $x(0) = x_0$).

The differential inclusion can take a more general form:

$$\begin{cases} x'(t) \in F(t, x(t)) \\ x(0) \in \Gamma \end{cases} \tag{4.67}$$

where $F : \mathbb{R} \times \mathbb{R}^n \to \mathscr{P}(\mathbb{R}^n)$ and $\Gamma \subset \mathbb{R}^n$. In this case, which we call the time-
dependent case, the selections of F that are measurable with respect to the time
variable are considered, and hence the concept of solution is also more general. It is
an absolutely continuous (see Appendix) function that satisfies the inclusion a.e.
in (4.67), obtained by solving the differential equation $x'(t) = f(t, x(t)), x(0) = x_0 \in \Gamma$, where f is a selection of F.

The absolutely continuous functions are the weakest acceptable solutions,
according to [3], since they are continuous and, moreover, they are differentiable
except on a set of measure zero. This allows solutions with discontinuities in its
derivatives at some points and at the same time avoids some bizarre cases (such as
a function that has derivative zero a.e. but is strictly monotonic).

An FDI is a generalization of a differential inclusion and was first proposed by
Aubin [2] and Baidosov [4]. It is symbolically written as

$$\begin{cases} x'(t) \in F(t, x(t)) \\ x(0) \in X_0 \end{cases} \tag{4.68}$$

and, as [18] proposed, is interpreted levelwise as the family of differential inclusions

$$\begin{cases} x'(t) \in [F(t, x(t))]_\alpha \\ x(0) \in [X_0]_\alpha \end{cases} \tag{4.69}$$

for all $\alpha \in [0, 1]$, where $[F]_\alpha : [0, T] \times \mathbb{R}^n \to \mathscr{K}_{\mathscr{C}}^n$ and $[X_0]_\alpha \in \mathscr{K}_{\mathscr{C}}^n$.

A solution to Problem (4.69) is an absolutely continuous function $x : [0, T] \to \mathbb{R}^n$
that satisfies the inclusion a.e. in $[0, T]$ and $x(0) = x_0 \in [X_0]_\alpha$. The set of all
solutions of (4.69) is denoted by $\Sigma_\alpha(x_0, T)$ and the attainable set at $t \in [0, T]$
by $\mathscr{A}_\alpha(x_0, t)$. Diamond [14] proved that the sets $\Sigma_\alpha(x_0, T)$ are the α-cuts of the
fuzzy solution $\Sigma(x_0, T)$ of (4.68), a fuzzy subset in $Z_T(\mathbb{R}^n)$ (see Appendix), that is,
$\Sigma(x_0, T) \in \mathscr{F}(Z_T(\mathbb{R}^n))$.

In the FDIs some trajectories may have more "preference" than the others which
is characterized by the value of its membership degree. This discrimination does not
exist in the traditional differential inclusions.

The following assumption assures that all the absolutely continuous solutions to
Problem (4.69) are defined on the same interval of existence.

Let Ω be an open subset in $\mathbb{R} \times \mathbb{R}^n$ such that $(0, x_0) \in \Omega$ and H a mapping from
Ω into the compact and convex subsets of $\mathbb{R}^n$. If there exist $b, T, M > 0$ such that:

- the set $Q = [0, T] \times (x_0 + (b + MT)B^n) \subset \Omega$, where B^n is the unit ball of $\mathbb{R}^n$;
- H maps Q into the ball of radius M

then it is said that the *boundedness assumption* holds (see [3, 14]).

The existence of solutions to FDIs for the time-dependent case is assured by Theorem 4.7, stated and proved in [14]. It does not require continuity of the differential field, but needs upper semicontinuity and boundedness assumption.

Theorem 4.7 ([14]). *Suppose that $X_0 \in \mathscr{F}_\mathscr{C}(\mathbb{R}^n)$, let Ω be an open set in $\mathbb{R} \times \mathbb{R}^n$ containing $\{0\} \times supp(X_0)$ and let $F : \Omega \times \mathscr{F}_\mathscr{C}(\mathbb{R}^n) \to \mathscr{F}_\mathscr{C}(\mathbb{R}^n)$ be usc. Suppose that the boundedness assumption holds for all $x_0 \in supp(X_0)$ and*

$$x' \in [F(t, x)]_0, \quad x(0) \in supp(X_0). \tag{4.70}$$

The families $\Sigma_\alpha(X_0, T)$ of all solutions to (4.69) are compact subsets in $Z_T(\mathbb{R}^n)$ for all $\alpha \in [0, 1]$. Moreover, these subsets are α-cuts of a fuzzy subset in $Z_T(\mathbb{R}^n)$, $\Sigma(X_0, T) \in \mathscr{F}_\mathscr{K}(Z_T(\mathbb{R}^n))$, which is the solution to (4.68). The attainable sets $\mathscr{A}_\alpha(X_0, t)$ of $\Sigma_\alpha(X_0, T)$ define the fuzzy subset $\mathscr{A}(X_0, t) \in \mathscr{F}_\mathscr{K}(\mathbb{R}^n)$.

The selection method is considered in order to search for solutions. It consists in finding a selection $f(t, x)$ of the set-valued function $[F(t, x)]_\alpha$ and solving the classical IVP

$$\begin{cases} x'(t) = f(t, x(t)) \\ x(0) = x_0 \end{cases}, \tag{4.71}$$

where $x_0 \in [X_0]_\alpha$. If f is continuous and bounded, for instance, there will be a solution to (4.71) and hence a solution to the differential inclusion.

Example 4.9. Let us solve the FDI associated with the family of problems

$$\begin{cases} x'(t) \in [-\lambda x(t)]_\alpha \\ x(0) = x_0 \in [X_0]_\alpha \end{cases}, \tag{4.72}$$

where $\lambda, x_0 \in \mathbb{R}^+$, $X_0 \in \mathscr{F}_\mathscr{C}(\mathbb{R})$ and $supp(X_0) \subset \mathbb{R}^+$. The function $[-\lambda x(t)]_\alpha$ is a singleton, therefore (4.72) is equivalent to

$$\begin{cases} x'(t) = -\lambda x(t) \\ x(0) = x_0 \in [X_0]_\alpha \end{cases}, \tag{4.73}$$

The set of all solutions is

$$\Sigma_\alpha(X_0, T) = \{x : x(t) = x_0 e^{-\lambda t}, x_0 \in [X_0]_\alpha\}. \tag{4.74}$$

These are the α-cuts of the fuzzy bunch of functions that is the solution to the FDI. Its attainable sets are

$$\mathscr{A}(X_0, t) = e^{-\lambda t} X_0. \tag{4.75}$$

Note that the FDI (4.72) is comparable with the decay model with initial condition in Example 4.5, since $F(t, X) = \lambda X$ is the extension of $f(t, x) = \lambda x$. Indeed, the attainable sets $\mathscr{A}(X_0, t)$ are the same as the solution at t via (ii)-differentiability, the one that presents decreasing diameter. And it is obviously different from the H-differentiable solution, that has increasing diameter (Figs. 4.17 and 4.18).

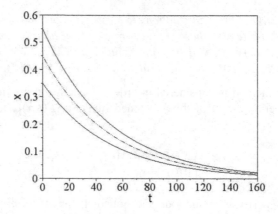

Fig. 4.17 Attainable sets of the 0-level (*continuous line*) and the core (*dashed-dotted line*) of solution to the decay model via FDIs in Example 4.9. Initial condition (0.35; 0.45; 0.55) and parameter $\lambda = 0.02$

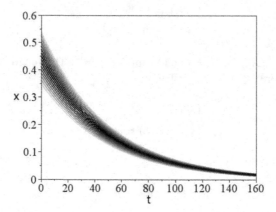

Fig. 4.18 Solution to the decay model via FDIs in Example 4.9, that is, a fuzzy bunch of functions whose membership of each function to the solution is represented by the scale of *gray*: the darker the color the higher the membership degree. Initial condition (0.35; 0.45; 0.55) and parameter $\lambda = 0.02$

Example 4.10. Now we also consider the parameter as a fuzzy number:

$$\begin{cases} x'(t) \in [-\Lambda x(t)]_\alpha \\ x(0) = x_0 \in [X_0]_\alpha \end{cases}, \qquad (4.76)$$

where $\Lambda \in \mathscr{F}_{\mathscr{C}}(\mathbb{R})$, $\mathrm{supp}(\Lambda) \subset \mathbb{R}^+$, $x_0 \in \mathbb{R}^+$, $X_0 \in \mathscr{F}_{\mathscr{C}}(\mathbb{R})$ and $\mathrm{supp}(X_0) \subset \mathbb{R}^+$. All solutions to (4.76) are confined between two values, for each t:

$$\begin{cases} x^-(t) = \min\{x(t) : x'(t) = \lambda(t)x(t), \ \lambda(t) \in [-\Lambda]_\alpha, \ x(0) \in [X_0]_\alpha\} \\ x^+(t) = \max\{x(t) : x'(t) = \lambda(t)x(t), \ \lambda(t) \in [-\Lambda]_\alpha, \ x(0) \in [X_0]_\alpha\} \end{cases}. \qquad (4.77)$$

That is,

$$\begin{cases} x^-(t) = x_{0\alpha}^- e^{-\lambda_\alpha^+ t} \\ x^+(t) = x_{0\alpha}^+ e^{-\lambda_\alpha^- t} \end{cases}. \qquad (4.78)$$

The set of all solutions has the attainable sets

$$\mathscr{A}_\alpha(X_0, t) = [x_{0\alpha}^- e^{-\lambda_\alpha^+ t}, x_{0\alpha}^+ e^{-\lambda_\alpha^- t}]. \qquad (4.79)$$

The FDI that has been just solved is also comparable to one FDE in the previous section, namely System (4.52). The present solution has decreasing diameter, hence it is clearly different from H-differentiable solution. It also has different attainable sets from the (ii)-differentiable solution. Example 4.6 showed that the (ii)-differentiable solution collapses at a certain value of t. The solution via differential inclusions is defined for all $t > 0$, preserving the property of asymptotic solution of the classical case (Fig. 4.19).

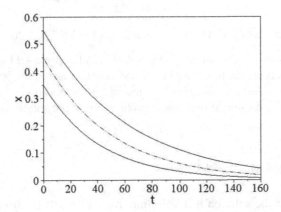

Fig. 4.19 Attainable sets of the 0-level (*continuous line*) and the core (*dashed-dotted line*) of solution to the decay model via FDIs in Example 4.10. Initial condition $(0.35; 0.45; 0.55)$ and parameter $\Lambda = (0.016; 0.020; 0.024)$

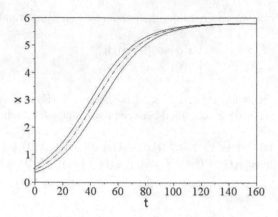

Fig. 4.20 Attainable sets of the 0-level (*continuous line*) and the core (*dashed-dotted line*) of solution to the logistic model via FDIs in Example 4.11. Initial condition (0.35; 0.45; 0.55) and parameters $k = 5.8$ and $a = 0.01$

Example 4.11. Let us solve the FDI associated with the family of problems

$$\begin{cases} x'(t) \in [a(k - x(t))]_\alpha \\ x(0) = x_0 \in [X_0]_\alpha \end{cases}, \tag{4.80}$$

where $a, k, x_0 \in \mathbb{R}^+$, and $X_0 \in \mathscr{F}_\mathscr{C}(\mathbb{R})$ and supp$(X_0) \subset \mathbb{R}^+$. As in Example 4.9, the function $[a(k - x(t))]_\alpha$ is a singleton. The set of all solutions of (4.80) is obtained by solving the classical case and varying the value of the initial condition (as in Example 4.9):

$$\Sigma_\alpha(X_0, T) = \left\{ x : x(t) = \frac{kx_0 e^{akt}}{k + x_0(e^{akt} - 1)}, x_0 \in [X_0]_\alpha \right\}. \tag{4.81}$$

The fuzzy solution $\Sigma(X_0, T)$ has α-cuts given by (4.81) (Figs. 4.20 and 4.21).

The functions that constitute the fuzzy solution in Example 4.11 are solutions of the associated IVP just as in Example 4.9 and hence preserve the properties of the classical case. In these two examples the method for finding solutions is the same as the one that will be presented in the next section, hence the solutions are the same.

4.5 Extension of the Solution

The extension of the solution is a very intuitive method. It consists in solving an ODE and extending the solution according to a fuzzy parameter, which can be the initial condition or some other parameter of the FDE. It will be clear that, besides

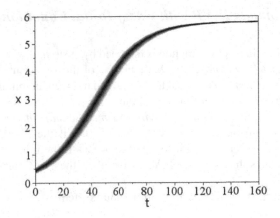

Fig. 4.21 Solution to the logistic model via FDIs in Example 4.11, that is, a fuzzy bunch of functions whose membership of each function to the solution is represented by the scale of *gray*: the darker the color the higher the membership degree. Initial condition $(0.35; 0.45; 0.55)$ and parameters $k = 5.8$ and $a = 0.01$

intuitive, the fuzzy solution preserves properties from the crisp solution. Hence the function may present decreasing length of its support, periodicity, and other behaviors that may be inherent of the phenomenon being modeled.

The method of extension of the solution was presented in [10, 26] for solving first-order FIVPs. Many other authors developed the same idea in the following years (see, for instance, [11, 16, 23]), though the authors do not always use the specific term *extension of the solution*.

It must be clear that this method does not solve an FDE and has no fuzzy derivatives involved. Solutions are obtained by extending the operator that associates each ODE and its parameter with a solution, *in each value in the domain*. The result of this operation is a fuzzy subset in $\mathbb{R}^n$. Buckley and Feuring [10] claims that it is equivalent to the united extension of the operator that associates each ODE and its parameter with a solution (obtaining a fuzzy subset in a space of functions) and calculating its attainable sets. This is a connection to the approach via fuzzification of the derivative operator (see Sect. 4.6).

We consider the IVP

$$\begin{cases} x'(t) = f(t, x(t), w) \\ x(0) = x_0 \end{cases}, \tag{4.82}$$

where $x_0 \in \mathbb{R}^n$, $w \in \mathbb{R}^k$ and $f : \mathbb{R} \times \mathbb{R}^n \to \mathbb{R}^n$ is continuous. A solution is a continuous function that satisfies the initial condition and the differential equation for all t. For each pair of parameter and initial condition there is a solution associated with (4.82) that we denote by $x(\cdot, x_0, w)$.

4.5.1   Autonomous FIVP with Fuzzy Initial Condition

Let us first consider that a phenomenon is modeled by system (4.82) and that only its initial condition is a fuzzy subset. Each element x_0 of this fuzzy subset X_0 leads to a different solution $x(\cdot, x_0)$. It is natural to consider that, given an initial value, there is a solution and, given a set of initial values, there is a set of solutions. *Each classical solution evaluated at t is associated with the membership of the correspondent initial value via extension principle.* This fuzzy subset is the solution to the FIVP at t. In other words, given $x(\cdot, x_0)$ a solution to (4.82), if the initial condition is the fuzzy subset X_0, the solution to the FIVP is based on the extension of $x(t, x_0)$, that is, $\hat{x}(t, X_0)$, $X_0 \in \mathscr{F}_{\mathscr{C}}(R^n)$.

The authors in [23] considered the autonomous system:

$$\begin{cases} x'(t) = f(x(t)) \\ x(0) = x_0 \end{cases}, \tag{4.83}$$

with fuzzy initial value. They have proved existence and uniqueness of the solution via extension of the solution and also demonstrated the equivalence with the FDI

$$\begin{cases} x'(t) = f(x(t)) \\ x(0) \in X_0 \end{cases}, \tag{4.84}$$

where $X_0 \in \mathscr{F}_{\mathscr{C}}(R^n)$.

Theorem 4.8 ([23]). *Consider U an open set in $\mathbb{R}^n$ and suppose that the IVP (4.83) admits only one solution $x(\cdot, x_0)$ for each $x_0 \in U$, with f continuous. Suppose also that $x(t, \cdot)$ depends continuously on the initial condition. Given $X_0 \in \mathscr{F}_{\mathscr{C}}(U)$, the extension $\hat{x}(t, X_0)$ of $x(t, X_0)$, that is, the solution of the FIVP with fuzzy initial condition correspondent to (4.83), is well defined. Moreover, the solution coincides with the (attainable sets of the) solution of the FDI (4.84).*

Remark 4.1. We can assure existence of the solution by using Lipschitz condition. That is, if there exists a constant $L > 0$ such that

$$||f(x) - f(y)|| \le L||x - y|| \tag{4.85}$$

then there is only one solution $x(\cdot, x_0)$, for each $x_0 \in \mathbb{R}^n$, to IVP (4.83).

Remark 4.2. Theorem 4.8 establishes that, given some conditions, the attainable sets of the solution via FDIs are the same as the solution calculated at each $t \in$ via extension of the solution.

Example 4.12. The classical decay model

$$\begin{cases} x'(t) = -\lambda x(t) \\ x(0) = x_0 \end{cases}, \tag{4.86}$$

where $\lambda \in \mathbb{R}^+$ and $x_0 \in \mathbb{R}^+$ has

$$x(t) = x_0 e^{-\lambda t} \qquad (4.87)$$

as solution.

According to the method of the present section, if the initial condition x_0 is considered to be fuzzy, that is, $x_0 = X_0 \in \mathscr{F}_\mathscr{C}(\mathbb{R})$, the solution to the new problem is the extension of the solution $x(t) = x(t, x_0)$ with respect to the initial condition. Levelwise we obtain

$$[X(t)]_\alpha = \hat{x}(t, X_0) = \{x_0 e^{-\lambda t}, x_0 \in [X_0]_\alpha\}. \qquad (4.88)$$

Example 4.13. The logistic model

$$\begin{cases} x'(t) = ax(t)(k - x(t)) \\ x(0) = x_0 \end{cases}, \qquad (4.89)$$

where $a, k, x_0 \in \mathbb{R}^+$ is known to have the solution

$$x(t) = \frac{kx_0 e^{akt}}{k + x_0(e^{akt} - 1)}. \qquad (4.90)$$

Considering the initial condition x_0 a fuzzy number X_0, the solution to the new problem is the extension of the solution $x(t) = x(t, x_0)$:

$$[X(t)]_\alpha = \hat{x}(t, X_0) = \left\{ \frac{kx_0 e^{akt}}{k + x_0(e^{akt} - 1)}, x_0 \in [X_0]_\alpha \right\}. \qquad (4.91)$$

4.5.2 FIVP with Fuzzy Initial Condition and Fuzzy Parameter

The general case, in which the FIVP is not autonomous and there are other fuzzy parameters influencing the differential equation, is treated by Bede [7]. If the parameter w is also fuzzy, the solution to the FIVP is the extension of $x(t, x_0, w)$, that is, $\hat{x}(t, X_0, W)$, $X_0 \in \mathscr{F}_\mathscr{C}(R^n)$, $W \in \mathscr{F}_\mathscr{C}(R^k)$. According to the result in [27], Lipschitz condition on the second and third variables of f assures existence, uniqueness, and continuity of the solution, with respect to the parameter w and the initial condition. Bede [7] uses this result to state the existence and uniqueness theorem for the FIVP.

Remark 4.3. Many authors write $X'(t) = \hat{F}(t, X(t), W), X(0) = X_0$, which is just a notation. As we said, this is not an FDE since there is no fuzzy derivative, hence $X'(t)$ does not make sense.

Theorem 4.9 ([7]). *Let $f : [t_0, t_0 + p] \times [x_0 - q, x_0 + q] \times \overline{B}(w, r)$ and assume that F is Lipschitz in its second and third variables, that is, there exist constants $L_1 > 0$ and $L_2 > 0$ such that*

$$\|f(t, x, w) - f(t, y, w)\| \leq L_1 \|x - y\| \tag{4.92}$$

and

$$\|f(t, x, w) - f(t, x, z)\| \leq L_2 \|w - z\| \tag{4.93}$$

Then the solution to (4.82) with fuzzy parameter, defined as the extension $\hat{x}(t, X_0, W)$, where $x(\cdot, x_0, w)$ is solution to (4.82) in its classical form, is well defined, unique, and continuous. Moreover, it can be defined levelwise:

$$[X(t)]_\alpha = [\hat{x}(t, X_0, W)]_\alpha = x(t, [X_0]_\alpha, [W]_\alpha). \tag{4.94}$$

Example 4.14. Now in the decay model we also consider the parameter λ as a fuzzy number. The solution is the extension of the solution $x(t) = x(t, x_0, \lambda)$ with respect to the x_0 and λ. Levelwise we obtain

$$[X(t)]_\alpha = \hat{x}(t, X_0, \Lambda) = \{x_0 e^{-\lambda t}, x_0 \in [X_0]_\alpha, \lambda \in [\Lambda]_\alpha\}. \tag{4.95}$$

Hence

$$[X(t)]_\alpha = [x_{0\alpha}^- e^{-\lambda_\alpha^+ t}, x_{0\alpha}^+ e^{-\lambda_\alpha^- t}]. \tag{4.96}$$

In this case, the solution is the same as via FDIs (see Example 4.10), but it does not happen in general (see [1]).

Example 4.15. Now consider the parameter k and a in the logistic model as fuzzy numbers K and A with $\mathrm{supp}(K) \subset \mathbb{R}^+$ and $\mathrm{supp}(A) \subset \mathbb{R}^+$ and set $Z = A \times K$. The solution is the extension of the solution $x(t) = x(t, x_0, z)$ with respect to x_0 and $z = k \times a$. Levelwise we obtain

$$[X(t)]_\alpha = \hat{x}(t, X_0, Z) = \left\{ \frac{kx_0 e^{akt}}{k + x_0(e^{akt} - 1)}, x_0 \in [X_0]_\alpha, a \in [A]_\alpha, k \in [K]_\alpha \right\}. \tag{4.97}$$

To obtain the expression of the level set functions we need to find the minimum and the maximum of the expression above. We calculate numerically the 0-levels and cores of the solution at each $t \in [0, T]$ and display it in Fig. 4.22.

The extension of the solution is an intuitive method. However, calculating the solution at each t in the domain is not obvious. It demands minimization and maximization for each t and each α-cut and, most of the times, it cannot be done analytically. The previous approaches present the same difficulty. The next approach

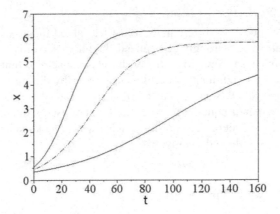

Fig. 4.22 The 0-level (*continuous line*) and the core (*dashed-dotted line*) of solution to the logistic model via extension of the solution in Example 4.15. Initial condition $(0.35; 0.45; 0.55)$ below carrying support $K = (5.3; 5.8; 6.3)$ and growth parameter $A = (0.005; 0.010; 0.015)$

does not solve this problem, since operating with fuzzy subsets is complex, which also makes the theory more challenging. On the contrary, the next approach unifies all the others presented so far.

4.6   Extension of the Derivative Operator

The $\hat{D}$-derivative operates on fuzzy subsets of functions, but the equality in the FDE is evaluated for each t, that is, on the attainable sets.

FIVP (1.2) becomes

$$\begin{cases} \hat{D}X(t) = F(t, X(t)) \\ X(0) \; = X_0 \end{cases}, \qquad (4.98)$$

where $X_0 \in \mathscr{F}(\mathbb{R}^n)$ and $\hat{D}X(t) \in \mathscr{F}(\mathbb{R}^n)$ is the attainable set of the $\hat{D}$-derivative (see Sect. 3.2.2) of the fuzzy bunch of functions $X(\cdot)$ at t.

This approach is not equivalent to any other considered, but has many similarities with them. Some points should be highlighted:

- $\hat{D}$ is a fuzzy derivative and so are Hukuhara and strongly generalized derivatives. FDIs and the extension of the solution do not use fuzzy derivatives.

(continued)

- $\hat{D}$-derivative does not differentiate fuzzy-set-valued functions (differently from Hukuhara and strongly generalized Hukuhara derivatives). It operates on fuzzy bunches of functions (fuzzy subsets in spaces of functions).
- Provided some conditions, $\hat{D}$-derivative operates by differentiating classical functions (as FDIs and extension of the solution).
- The FIVP demands equality between the fuzzy subset of the left-hand-side and the fuzzy subset of the right-hand-side, as with Hukuhara and strongly generalized Hukuhara derivatives.

In what follows we will prove that

1. Provided some hypotheses hold, the solution of the FIVP via FDIs is one solution via extension of the derivative operator.
2. Provided some hypotheses hold, the solution of the FIVP via extension of the solution is one solution via extension of the derivative operator (in the sense of attainable sets).
3. Provided some conditions hold, the solutions of the FIVP via strongly generalized Hukuhara derivative (and particularly, via Hukuhara derivative) are solutions via extension of the derivative operator (in the sense of attainable sets).

Items 1 and 2 will be briefly illustrated in the next example.

Example 4.16. The decay model with nonfuzzy coefficient was solved using other methods. Now consider

$$\begin{cases} \hat{D}X(t) = -\lambda X(t) \\ X(0) = X_0 \end{cases}, \tag{4.99}$$

where $\lambda \in \mathbb{R}^+$, $X_0 \in \mathscr{F}_{\mathscr{C}}(\mathbb{R})$ and $\text{supp}(X_0) \subset \mathbb{R}^+$.

The solution via FDIs,

$$\Sigma_\alpha(X_0, T) = \{x : x(t) = x_0 e^{-\lambda t}, x_0 \in [X_0]_\alpha\}. \tag{4.100}$$

is solution to (4.99), since $X(\cdot) = \Sigma(X_0, T)$ satisfies the hypothesis of Theorem 3.13 and hence

$$\begin{aligned} [\hat{D}X(\cdot)]_\alpha &= D[X(\cdot)]_\alpha \\ &= \{Dx(\cdot) : x(t) = x_0 e^{-\lambda t}, x_0 \in [X_0]_\alpha\} \\ &= \{\lambda x(\cdot) : x(t) = x_0 e^{-\lambda t}, x_0 \in [X_0]_\alpha\} \\ &= [\lambda X(\cdot)]_\alpha \end{aligned} \tag{4.101}$$

for all $\alpha \in [0, 1]$.

Since it has already been shown that the attainable sets of the cited solution is solution via extension of the solution, the FIVP via extension of the solution and FDIs have the same solution, which is one solution via $\hat{D}$-derivative.

The fact that the solutions via FDIs and via $\hat{D}$-derivative are the same is not a mere coincidence. In what follows we make this connection between these two theories. Let us first recall the FIVP modeled by FDIs:

$$\begin{cases} x'(t) \in F(t, x(t)) \\ x(0) \in X_0 \end{cases} \tag{4.102}$$

Levelwise it is equivalent to

$$\begin{cases} x'_\alpha(t) \in [F(t, x_\alpha(t))]_\alpha \\ x_\alpha(0) \in [X_0]_\alpha \end{cases} . \tag{4.103}$$

Taking the union of all functions $x_\alpha(\cdot)$

$$\begin{cases} \bigcup x'_\alpha(t) \subseteq \bigcup [F(t, x_\alpha(t))]_\alpha \\ \bigcup x_\alpha(0) \subseteq [X_0]_\alpha \end{cases} . \tag{4.104}$$

The union of all solutions $x_\alpha(\cdot)$ of the differential inclusion (4.103) defines the α-cut of the solution $X(\cdot)$ of problem (4.102). We can rewrite last system as

$$\begin{cases} D[X(t)]_\alpha \subseteq \bigcup [F(t, x_\alpha(t))]_\alpha \\ [X(0)]_\alpha \subseteq [X_0]_\alpha \end{cases} , \tag{4.105}$$

where $[X(\cdot)]_\alpha = \{x_\alpha(\cdot) : x_\alpha(\cdot) \text{ is solution to (4.103)}\}$.

We have $[X(0)]_\alpha = [X_0]_\alpha$ by the construction of the solution. We are interested in finding conditions for

$$\left[\hat{D}X(t)\right]_\alpha = [F(t, X(t))]_\alpha \tag{4.106}$$

to hold, that is,

$$\left[\hat{D}X(t)\right]_\alpha \subseteq [F(t, X(t))]_\alpha \tag{4.107}$$

and

$$\left[\hat{D}X(t)\right]_\alpha \supseteq [F(t, X(t))]_\alpha . \tag{4.108}$$

If $D[X(t)]_\alpha = \left[\hat{D}X(t)\right]_\alpha$, the condition

$$[F(t, X(t))]_\alpha = \bigcup [F(t, x_\alpha(t))]_\alpha \tag{4.109}$$

guarantees (4.107) Since $X(\cdot)$ is a solution via FDIs, it has compact α-cuts in $Z_T(\mathbb{R})$ and hence Theorem 3.13 can be used.

Hence the solution via FDIs is a good candidate for being a solution to (4.98), provided condition (4.109) holds. We need only to prove (4.108).

Example 4.17. The function $F(t, X(t)) = \lambda X(t)$, where $\lambda \in \mathbb{R}$, satisfies condition (4.109), that is,

$$\cup_{x_\alpha \in [X]^\alpha} [\lambda x_\alpha(t)]^\alpha = [\lambda X(t))]^\alpha. \tag{4.110}$$

If the parameter Λ is fuzzy, we also have that:

$$\cup_{x_\alpha \in [X]^\alpha} [\Lambda x_\alpha(t)]^\alpha = [\Lambda X(t))]^\alpha. \tag{4.111}$$

Note that λX is the extension of λx with respect to x according to Definition 2.5. And according to Definition 2.6, ΛX is the extension of Λx. On the other hand, $(1 - X)X$ is not the extension of $(1 - x)x$ and

$$\cup_{x_\alpha \in [X]^\alpha} [(1 - x_\alpha(t))x_\alpha(t)]^\alpha \subset [(1 - X(t))X(t)]^\alpha. \tag{4.112}$$

The next result is important to prove the existence Theorem 4.11.

Theorem 4.10 (Michael's Selection Theorem, See e.g. [3]). *Let $\mathbb{X}$ be a metric space, $\mathbb{Y}$ a Banach space, and G a map from $\mathbb{X}$ to convex and closed subsets of $\mathbb{Y}$. If G is lower semicontinuous, then there exists a continuous selection $f : \mathbb{X} \to \mathbb{Y}$ of G.*

The following statement is a result of Michael's Selection Theorem according to [3], p. 83: if $\mathbb{X}$ is a paracompact space ($\mathbb{R}^n$ is paracompact), for any $y_0 \in G(x_0)$ the set-valued map G_0 defined by

$$G_0(x_0) = \{y_0\}, \quad G_0(x) = G(x) \quad \forall x \neq x_0 \tag{4.113}$$

is also lsc with convex values and hence there exists a continuous selection g_0 of G_0. In other words, *for every y_0 in $G(x_0)$ there passes a continuous selection of G.*

The proof of the existence Theorem 4.11 (published in [5]) is reproduced next.

Theorem 4.11 ([5]). *Let $X_0 \in \mathscr{F}_C(\mathbb{R}^n)$ and Ω be an open set in $\mathbb{R} \times \mathbb{R}^n$ containing $\{0\} \times suppX_0$ and $F : \mathbb{R} \times \mathscr{F}_{\mathscr{K}}(\mathbb{R}^n) \to \mathscr{F}_{\mathscr{K}}(\mathbb{R}^n)$ a fuzzy-set-valued function such that $F(t, x) = F|_\Omega$ is continuous with $[F(t, x)]^\alpha$ compact, convex, and $[F(t, X)]^\alpha = \bigcup_{x \in [X]^\alpha} [F(t, x)]^\alpha$. Also, suppose that the boundedness assumption holds. Then, there exists a solution $X(\cdot) \in \mathscr{F}_{\mathscr{K}}(Z_T(\mathbb{R}^n))$ for problem (4.98). Moreover, $[X(t)]^\alpha$ are compact and connected in $\mathbb{R}^n$, for all $\alpha \in [0, 1]$.*

Proof. The stated hypotheses are stronger than those in Theorem 4.7, including condition (4.109). Hence the solution $X(\cdot)$ to (4.102), whose existence is guaranteed, satisfies (4.107). In what follows we prove that $X(\cdot)$ also satisfies (4.108).

We want to prove that given $z \in [F(s, x)]_\alpha$ there exists $x(\cdot) \in [X(\cdot)]_\alpha$ such that $x'(s) = z$, or $z \in D[X]_\alpha$. Note that, given $z \in [F(s, x)]_\alpha$, there exists $y(\cdot) \in [X(\cdot)]_\alpha$ such that $y(s) = x$, hence $z \in [F(s, y(s))]_\alpha$. Michael's Selection Theorem applied to $[F(t, y(t))]_\alpha$ guarantees that there exists a continuous selection $z(\cdot)$ such that $z(t) \in [F(t, y(t))]_\alpha$, for all $t \in [0, T]$.

We use this selection $z(\cdot)$ to define

$$x(t) = x_0 + \int_0^t z(\tau)d\tau \tag{4.114}$$

for some $x_0 \in [X_0]_\alpha$.

This function $x(\cdot)$ belongs to $[X(\cdot)]_\alpha$, since $x'(t) = z(t)$ for all $t \in [0, T]$. Hence we have proved (4.108). ∎

Remark 4.4. Theorem 4.11 establishes that, provided some conditions hold, the solution via FDIs is the same as via $\hat{D}$-derivative, that is, the fuzzy bunches of functions that satisfy the respective FIVPs are the same. Consequently, the attainable sets obtained from the two methods are the same.

FIVPs using extension of real-valued functions has a rich literature (see [11, 13, 21, 23]). The generalized case, though, is not much mentioned. In what follows we use the second case, that is, $f : \mathbb{R}^n \to \mathscr{F}(\mathbb{R}^n)$, which has a wider application.

Corollary 4.1. *Let $X_0 \in \mathscr{F}_C(R)$, Ω be an open set in $\mathbb{R} \times \mathbb{R}^n$ containing $\{0\} \times \text{supp}X_0$, $f : \mathbb{R} \times \mathbb{R} \to \mathscr{F}_\mathscr{C}(\mathbb{R})$ be d_∞-continuous and $\hat{f} : \mathbb{R} \times \mathscr{F}_\mathscr{C}(\mathbb{R}) \to \mathscr{F}_\mathscr{C}(\mathbb{R})$ be the extension of f. Also, let the boundedness assumption hold. Then the FIVP (4.98), with right-hand-side function $\hat{f}$, has a solution.*

Proof. From Theorem 2.9 $\hat{f}$ is d_∞-continuous and $[\hat{f}(t, X)]^\alpha = \bigcup_{x \in [X]^\alpha} [f(t, x)]^\alpha$. Thus the conditions of Theorem 4.11 are satisfied and therefore the FIVP (4.98) has a solution. ∎

Example 4.18. The FIVP

$$\begin{cases} \hat{D}X(t) = -\Lambda X(t) \\ X(0) = X_0 \end{cases}, \tag{4.115}$$

where $\Lambda \in \mathscr{F}_\mathscr{C}(\mathbb{R})$, $\text{supp}(\Lambda) \subset \mathbb{R}^+$, $X_0 \in \mathscr{F}_\mathscr{C}(\mathbb{R})$ and $\text{supp}(X_0) \subset \mathbb{R}^+$, has right-hand-side function given by the extension principle, according to Definition 2.6. Indeed, $F(X) = -\Lambda X$ is the extension of $G(x) = \Lambda x$.

We apply Theorem 4.11 and the solution via FDIs of Example 4.10 is solution to (4.115). Hence the attainable sets of the solution are

$$X(t) = [x_{0\alpha}^- e^{-\lambda_\alpha^+ t}, x_{0\alpha}^+ e^{-\lambda_\alpha^- t}]. \tag{4.116}$$

Reference [23] has proved that the solution of an IVP with fuzzy initial condition via extension of the solution is the same as the solution of the associated FDI (in terms of attainable sets), provided the function f is continuous and the IVP has a unique solution. As a result of Theorem 4.11 and Corollary 4.1, if the right-hand-side function F is the extension of a continuous function $f : \mathbb{R} \times \mathbb{R} \to \mathbb{R}$, the solution of the FDI is a solution to the associated FIVP with the derivative via extension principle. Hence, if the associated IVP has a unique solution for each initial condition, we have the same attainable sets of the fuzzy solutions for the three mentioned approaches.

We next prove the following autonomous case: if the differential field F is the extension of $f : \mathbb{R}^n \to \mathbb{R}^n$, the solutions via FDI, the extension of the solution to

$$\begin{cases} x'(t) = f(x(t)) \\ x(0) = x_0 \end{cases}, \tag{4.117}$$

and $\hat{D}$-derivative are the same (in terms of attainable sets).

Theorem 4.12. *Consider the FIVP*

$$\begin{cases} \hat{D}X(t) = \hat{f}(X(t)) \\ X(0) = X_0 \end{cases}, \tag{4.118}$$

where $X_0 \in \mathscr{F}_C(\mathbb{R}^n)$, $\hat{f}$ is the extension of a continuous function $f : \mathbb{R}^n \to \mathbb{R}^n$ such that (4.117) has only one solution. Then the solution of (4.118) is given by the extension of the solution of (4.117), $X(\cdot) = \hat{x}(\cdot, x_0)$.

Proof. The extension of the solution was previously (Sect. 4.5) defined at each $t \in [0, T]$. However, it is also possible to define it as a fuzzy bunch of functions, considering the solution not at each t, but as the whole function. References [10, 17] claim that both approaches are equivalent in terms of attainable sets. The approach we adopt here is the one that deals with fuzzy bunch of functions.

Consider $x_\alpha(\cdot, x_0)$ an element of the α-cut of the extension $X(\cdot) = \hat{x}(\cdot, x_0)$ of the solution $x(\cdot, x_0)$ to (4.117).

$$\begin{aligned} Dx_\alpha(\cdot) &= f(x_\alpha(\cdot)) \\ \bigcup_{x_\alpha \in [X(\cdot)]_\alpha} Dx_\alpha(\cdot) &= \bigcup_{x_\alpha \in [X(\cdot)]_\alpha} f(x_\alpha(\cdot)) \\ D\bigcup_{x_\alpha \in [X(\cdot)]_\alpha} x_\alpha(\cdot) &= f(\bigcup_{x_\alpha \in [X(\cdot)]_\alpha} x_\alpha(\cdot)) \\ D[X(\cdot)]_\alpha &= f([X(\cdot)]_\alpha) \end{aligned} \tag{4.119}$$

Since $X(\cdot)$ is also solution (fuzzy bunch of functions) via FDIs, we have $X(\cdot) \in \mathscr{F}_{\mathscr{H}}(Z_T(\mathbb{R}^n))$ and Theorem 3.13 is valid. Also, since f is continuous, Theorem 2.5 holds. Hence

$$\left[\hat{D}X(\cdot)\right]_\alpha = D[X(\cdot)]_\alpha = f([X(\cdot)]_\alpha) = [\hat{f}(X(\cdot))]_\alpha \tag{4.120}$$

for all $\alpha \in [0, 1]$. Thus

$$\hat{D}X(\cdot) = \hat{f}(X(\cdot)) \tag{4.121}$$

and, in particular,

$$\hat{D}X(t) = \hat{f}(X(t)) \tag{4.122}$$

for all $t \in [0, T]$.

Remark 4.5. Theorem 4.12 establishes that, under some conditions, the fuzzy bunches of functions of the solution via extension of the solution is the same as via $\hat{D}$-derivative. Consequently, the attainable sets obtained from the two methods are the same.

Example 4.19. The solution of the logistic model

$$\begin{cases} \hat{D}X(t) = \hat{f}(t, X(t))) \\ X(0) = X_0 \end{cases}, \tag{4.123}$$

where $f(t, x) = ax(k - x)$, $a, k \in \mathbb{R}^+$, $X_0 \in \mathscr{F}_{\mathscr{C}}(\mathbb{R})$ and supp$(X_0) \subset \mathbb{R}^+$ is based on the solution of Example 4.13, that is, the same problem solved via extension of the solution. In that example, we solved the classical differential equation. Instead of extending the solution at each t of the domain, here we extend the solutions as elements in the space of functions. Hence we consider the solution $x(\cdot)$ such that

$$x(t) = \frac{kx_0 e^{akt}}{k + x_0(e^{akt} - 1)}. \tag{4.124}$$

Applying Theorem 4.123, a solution to (4.123) is the extension of $x(\cdot) = x(\cdot, x_0)$,

$$X(\cdot) = \hat{x}(\cdot, X_0). \tag{4.125}$$

As in Example 4.16, it can be verified that the solution satisfies (4.123) by applying Theorem 3.13:

$$\begin{aligned}
[\hat{D}X]_\alpha &= D[X]_\alpha \\
&= \left\{ Dx(\cdot) : x(t) = \frac{kx_0 e^{akt}}{k + x_0(e^{akt} - 1)}, x_0 \in [X_0]_\alpha \right\} \\
&= \left\{ ax(\cdot)(k - x(\cdot)) : x(t) = \frac{kx_0 e^{akt}}{k + x_0(e^{akt} - 1)}, x_0 \in [X_0]_\alpha \right\} \\
&= [\hat{f}(t, X(t))]_\alpha.
\end{aligned} \tag{4.126}$$

See Fig. 4.23 for illustration.

Fig. 4.23 Solution to the logistic model via $\hat{D}$-derivative in Example 4.19, that is, a fuzzy bunch of functions whose membership of each function to the solution is represented by the scale of *gray*: the darker the color the higher the membership degree. Initial condition $(0.35; 0.45; 0.55)$ and parameters $k = 5.8$ and $a = 0.01$

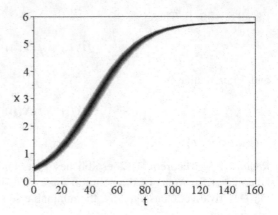

Note that we do not have the expression for $\hat{f}$ in terms of the standard arithmetic. As it has been already mentioned, $\hat{f}(t, X(t))) \not\subseteq aX(t)(k - X(t))$, hence

$$\begin{cases} \hat{D}X(t) = aX(t)(k - X(t)) \\ X(0) = X_0 \end{cases}, \qquad (4.127)$$

is a different problem, which will be solved using other method (see Theorem 4.13 and Example 4.20).

A connection between the generalized differentiabilities and the $\hat{D}$-derivative was established in Sect. 3.3. Since they coincide for the representative bunches of functions of a special class of fuzzy-number-valued numbers (see Theorems 3.17 and 3.18), it is natural to wonder if the solution to an FIVP also coincides in both approaches. Indeed, provided some conditions hold, they do: the hypotheses of the characterization Theorem 4.6 for strongly generalized differentiability assure two solutions (fuzzy-number-valued functions) that generate two different fuzzy bunches of functions that are solutions to the corresponding FIVP with $\hat{D}$-derivative.

Theorem 4.13. *Assume the hypotheses of Theorem 4.6 hold true for F and X_0 in FIVP (4.98) and that the solutions obtained from (4.47) and (4.48) belong to $\mathscr{F}_{\mathscr{C}}^0(\mathbb{R})$. Then FIVP (4.98) has at least two solutions.*

Proof. Theorem 4.6 assures two solutions via strongly generalized differentiability. We will prove that if the solutions assume values in $\mathscr{F}_{\mathscr{C}}^0(\mathbb{R})$, they satisfy Theorem 3.17 which provides us with two representative bunches of functions whose derivative is the same as those via generalized Hukuhara differentiability.

Let X be the solution to the FIVP via strongly generalized differentiability obtained by solving (4.47). It is obvious that X is continuous and x_α^- and x_α^+ are differentiable with respect to t. We will prove that this differentiability is uniform with respect to $\alpha \in [0, 1]$. Since X is strongly generalized differentiable, it is gH-differentiable (the latter is more general than the former) and, by Theorem 3.6, it satisfies (a) or (b) of Theorem 3.17.

First we will show uniform differentiability, that is, given $\epsilon > 0$, there exists $\delta > 0$ such that

$$\left| \frac{x_\alpha^\pm(t+h) - x_\alpha^\pm(t)}{h} - (x_\alpha^\pm)'(t) \right| < \epsilon \qquad (4.128)$$

if $|h| < \delta$, for all $\alpha \in [0, 1]$. In what follows we will use the fact that x_α^- and x_α^+ are solutions to (4.47) and thus satisfy

$$x_\alpha^\pm(t+h) = x_\alpha^\pm(t) + \int_t^{t+h} f_\alpha^\pm(s, x_\alpha^-(s), x_\alpha^+(s)) ds. \qquad (4.129)$$

Therefore,

$$
\begin{aligned}
\left| \frac{x_\alpha^\pm(t+h) - x_\alpha^\pm(t)}{h} - (x_\alpha^\pm)'(t) \right| &= \\
&= \left| \frac{x_\alpha^\pm(t+h) - x_\alpha^\pm(t)}{h} - f_\alpha^\pm(t, x_\alpha^-(t), x_\alpha^+(t)) \right| \\
&= \left| \frac{x_\alpha^\pm(t+h) - x_\alpha^\pm(t) - h f_\alpha^\pm(t, x_\alpha^-(t), x_\alpha^+(t))}{h} \right| \\
&= \left| \frac{x_\alpha^\pm(t) + \int_t^{t+h} f_\alpha^\pm(s, x_\alpha^-(s), x_\alpha^+(s)) ds - x_\alpha^\pm(t) - \int_t^{t+h} f_\alpha^\pm(t, x_\alpha^-(t), x_\alpha^+(t)) ds}{h} \right| \\
&\leq \frac{\int_t^{t+h} \left| f_\alpha^\pm(s, x_\alpha^-(s), x_\alpha^+(s)) - f_\alpha^\pm(t, x_\alpha^-(t), x_\alpha^+(t)) \right| ds}{|h|}
\end{aligned}
\qquad (4.130)
$$

The hypothesis of equicontinuity ensures that there is a $\delta > 0$ such that

$$\left| f_\alpha^\pm(s, x_\alpha^-(s), x_\alpha^+(s)) - f_\alpha^\pm(t, x_\alpha^-(t), x_\alpha^+(t)) \right| < \epsilon \text{ if } |s - t| < \delta, \qquad (4.131)$$

that is,

$$\int_t^{t+h} \left| f_\alpha^\pm(s, x_\alpha^-(s), x_\alpha^+(s)) - f_\alpha^\pm(t, x_\alpha^-(t), x_\alpha^+(t)) ds \right| < \epsilon |h|. \qquad (4.132)$$

This means that

$$\left| \frac{x_\alpha^\pm(t+h) - x_\alpha^\pm(t)}{h} - (x_\alpha^\pm)'(t) \right| < \epsilon. \qquad (4.133)$$

if $|h| < \delta$, that is, x_α^- and x_α^+ are differentiable real-valued functions with respect to t, uniformly with respect to α.

We have proved that the solution X obtained by solving (4.47) satisfies the conditions of Theorem 3.17. Hence the representative bunch of first kind, $\tilde{X}(\cdot)$ has $\hat{D}$-derivative equal to generalized Hukuhara derivative (in terms of attainable sets), which is the same as strongly generalized differentiability, since both exist. Thus

$$F(t, x(t)) = X'_G(t) = X'_{gH}(t) = \hat{D}\tilde{X}(t) \tag{4.134}$$

and it is proved that $\tilde{X}$ is a solution to FIVP (4.98).

Following the same reasoning one obtains that the solution to (4.48) leads to other representative bunch of first kind which is also solution to (4.98).

Example 4.20. All examples in Sect. 4.3, that is, Examples 4.5–4.8 have solutions X with $X(t) \in \mathscr{F}^0_{\mathscr{C}}(\mathbb{R})$. Hence, the representative bunches of first kind are solutions of the respective problems using the $\hat{D}$-derivative.

The last example of this section regards a system in which the state variable has values in $\mathbb{R}^2$. The solution belongs to $\mathscr{F}_K(\mathscr{A}C([0, T]; \mathbb{R}^2))$.

Example 4.21. Consider

$$\begin{cases} \hat{D}X(t) = F(X(t))) \\ X(0) = X_0 \end{cases}, \tag{4.135}$$

where $X_0 \in \mathscr{F}(\mathbb{R}^2)$ and $F : \mathscr{F}(\mathbb{R}^2) \to \mathscr{F}(\mathbb{R}^2)$ such that

$$\mu_{F(X)}(z, y) = \mu_X(y, z) \tag{4.136}$$

We will try the solution obtained via extension of the following associated problem:

$$\begin{cases} y'(t) = z(t), \ y(0) = y_0 \\ z'(t) = y(t), \ z(0) = z_0 \end{cases}, \tag{4.137}$$

whose solution is

$$x(t, x_0) = \begin{pmatrix} y(t) \\ z(t) \end{pmatrix} = \frac{y_0}{2} \left(\frac{e^t + e^{-t}}{e^t - e^{-t}} \right) + \frac{z_0}{2} \left(\frac{e^t - e^{-t}}{e^t + e^{-t}} \right) \tag{4.138}$$

where $x_0 = \begin{pmatrix} y_0 \\ z_0 \end{pmatrix}$.

The function $x(t, x_0)$ is continuous with respect to the initial condition. Hence the extension of this solution can be defined levelwise

$$[\hat{x}(t, X_0)]_\alpha = \hat{x}(t, [X_0]_\alpha) = \{x(t, x_0), x_0 \in [X_0]_\alpha\} \tag{4.139}$$

Hence

$$\left[\hat{D}\hat{x}(\cdot, X_0)\right]_\alpha = D\{x(\cdot) : x(t) = x(t, x_0), x_0 \in [X_0]_\alpha\}$$
$$= \{x'(\cdot) : x(t) = x(t, x_0), x_0 \in [X_0]_\alpha\} \qquad (4.140)$$

where

$$x'(t, x_0) = \begin{pmatrix} y'(t, y_0, z_0) \\ z'(t, y_0, z_0) \end{pmatrix} = \frac{y_0}{2}\left(\frac{e^t - e^{-t}}{e^t + e^{-t}}\right) + \frac{z_0}{2}\left(\frac{e^t + e^{-t}}{e^t - e^{-t}}\right)$$
$$= \begin{pmatrix} z(t, y_0, z_0) \\ y(t, y_0, z_0) \end{pmatrix} \qquad (4.141)$$

Thus,

$$\left[\hat{D}\hat{x}(t, X_0)\right]_\alpha = \{(z(t), y(t)) : x(t, x_0) = (y(t), z(t)), x_0 \in [X_0]_\alpha\}$$
$$= [F(\hat{x}(t, X_0))]_\alpha . \qquad (4.142)$$

The method that has been used to solve the last example is based on the extension of the solution. We have also shown that other techniques to find solutions to FIVPs involving $\hat{D}$-derivative are solving the FIVP via strongly generalized derivative and taking the representative bunches as solutions via $\hat{D}$-derivative; and using FDIs.

4.7 Summary

The most known approaches for FDEs were reviewed and compared to the new proposal developed in [5]. Some of the results are new in the literature and they are summarized next, together with other important concepts and results already known.

- The most known approaches for FDEs are via Hukuhara and strongly generalized derivatives (the former being a particular case of the latter), FDIs and extension of the solution.
- FDIs and the extension of the solution do not make use of fuzzy derivatives.
- FDEs via $\hat{D}$-derivative, introduced in [5] and herein further developed make use of fuzzy derivatives on fuzzy bunches of functions. Although it is considered a fuzzy derivative in this text, it differentiates classical functions as FDIs and the extension of the solution do.
- Conditions assure existence of solutions to FIVPs for all mentioned approaches.
- The use of attainable sets is necessary to compare the different approaches, since the solution of FIVPs via FDIs and $\hat{D}$-derivative are fuzzy bunches of functions, not fuzzy-set-valued functions as in the other proposals.

- The solutions of certain FIVPs via extension of the solution are the same as via FDIs.
- The solutions of certain FIVPs via $\hat{D}$-derivative are the same as via FDIs.
- The solutions of certain FIVPs via $\hat{D}$-derivative are the same as via extension of the solution.
- In terms of attainable sets the solutions of certain FIVPs via $\hat{D}$-derivative are the same as via strongly generalized derivative, that is, all the mentioned approaches can be reproduced via $\hat{D}$-derivative.

References

1. T. Allahviranloo, M. Shafiee, Y. Nejatbakhsh, A note on "fuzzy differential equations and the extension principle". Inf. Sci. **179**, 2049–2051 (2009)
2. J.P. Aubin, Fuzzy differential inclusions. Probl. Control Inf. Theory **19**(1), 55–67 (1990)
3. J.P. Aubin, A. Cellina, *Differential Inclusions: Set-Valued Maps and a Viability Theory* (Springer, Berlin/Heidelberg, 1984)
4. V.A. Baidosov, Fuzzy differential inclusions. PMM USSR **54**(1), 8–13 (1990)
5. L.C. Barros, L.T. Gomes, P.A. Tonelli, Fuzzy differential equations: an approach via fuzzification of the derivative operator. Fuzzy Sets Syst. **230**, 39–52 (2013)
6. B. Bede, Note on "numerical solutions of fuzzy differential equations by predictor-corrector method". Inf. Sci. **178**, 1917–1922 (2008)
7. B. Bede, *Mathematics of Fuzzy Sets and Fuzzy Logic* (Springer, Berlin/Heidelberg, 2013)
8. B. Bede, S.G. Gal, Generalizations of the differentiability of fuzzy-number-valued functions with applications to fuzzy differential equations. Fuzzy Sets Syst. **151**, 581–599 (2005)
9. B. Bede, S.G. Gal, Solutions of fuzzy differential equations based on generalized differentiability. Commun. Math. Anal. **9**, 22–41 (2010)
10. J.J. Buckley, T. Feuring, Almost periodic fuzzy-number-valued functions. Fuzzy Sets Syst. **110**, 43–54 (2000)
11. M.S. Cecconello, Sistemas dinâmicos em espaços métricos fuzzy – aplicações em biomatemática (in Portuguese). Ph.D. thesis, IMECC – UNICAMP, Campinas, 2010
12. Y. Chalco-Cano, H. Román-Flores, Some remarks on fuzzy differential equations via differential inclusions. Fuzzy Sets Syst. **230**, 3–20 (2013)
13. Y. Chalco-Cano, W.A. Lodwick, B. Bede, Fuzzy differential equations and Zadeh's extension principle, in *2011 Annual Meeting of the North American Fuzzy Information Processing Society (NAFIPS)*, 2011, pp. 1–5
14. P. Diamond, Time-dependent differential inclusions, cocycle attractors an fuzzy differential equations. IEEE Trans. Fuzzy Syst. **7**, 734–740 (1999)
15. L. Edelstein-Keshet, *Mathematical Models in Biology* (Society for Industrial and Applied Mathematics, Philadelphia, 2005)
16. N.A. Gasilov, I.F. Hashimoglu, S.E. Amrahov, A.G. Fatullayev, A new approach to non-homogeneous fuzzy initial value problem. Comput. Model. Eng. Sci. **85**, 367–378 (2012)
17. N.A. Gasilov, A.G. Fatullayev, S.E. Amrahov, A. Khastan, A new approach to fuzzy initial value problem. Soft Comput. **18**, 217–225 (2014)
18. E. Hüllermeier, An approach to modelling and simulation of uncertain dynamical systems. Int. J. Uncertainty Fuzziness Knowledge Based Syst. **5**(2), 117–137 (1997)
19. O. Kaleva, Fuzzy differential equations. Fuzzy Sets Syst. **24**, 301–317 (1987)
20. O. Kaleva, The Cauchy problem for fuzzy differential equations. Fuzzy Sets Syst. **35**, 389–396 (1990)
21. O. Kaleva, A note on fuzzy differential equations. Nonlinear Anal. **64**, 895–900 (2006)

22. A. Kandel, W.J. Byatt, Fuzzy processes. Fuzzy Sets Syst. **4**, 117–152 (1980)
23. M.T. Mizukoshi, L.C. Barros, Y. Chalco-Cano, H. Román-Flores, R.C. Bassanezi, Fuzzy differential equations and the extension principle. Inf. Sci. **177**, 3627–3635 (2007)
24. M. Mizumoto, K. Tanaka, The four operations of arithmetic on fuzzy numbers. Syst. Comput. Control **7**, 73–81 (1976)
25. J.J. Nieto, The Cauchy problem for continuous fuzzy differential equations. Fuzzy Sets Syst. **102**, 259–262 (1999)
26. M. Oberguggenberger, S. Pittschmann, Differential equations with fuzzy parameters. Math. Mod. Syst. **5**, 181–202 (1999)
27. L. Perko, *Differential Equations and Dynamical Systems* (Springer, New York, 2001)
28. S. Seikkala, On the fuzzy initial value problem. Fuzzy Sets Syst. **24**, 309–330 (1987)

Appendix A
Mathematical Background

Some readers may find it necessary to review some mathematical concepts, which we intend to briefly cover in this appendix. For further understanding refer to [1–4].

A.1 Continuity and Semicontinuity

Definition A.1. A function $f : X \to \mathbb{R}$ is said to be upper semi-continuous at x_0 if for any $\epsilon > 0$ there exist a $\delta > 0$ such that $f(x) < f(x_0) + \epsilon$ whenever $|x - x_0| < \delta$.

Definition A.2. A function $f : X \to \mathbb{R}$ is said to be lower semi-continuous at x_0 if for any $\epsilon > 0$ there exist a $\delta > 0$ such that $f(x) > f(x_0) - \epsilon$ whenever $|x - x_0| < \delta$.

Definition A.3. A family of real-valued functions $\{f_\alpha\}_\alpha$ is equicontinuous if given $\epsilon > 0$ and x_0 there exists $\delta > 0$ such that

$$|f_\alpha(x) - f_\alpha(x_0)| < \epsilon$$

whenever $|x - x_0| < \delta$, for all f_α.

A.2 Spaces of Functions

Denote by $L^0(\Omega)$ the space of all Lebesgue measurable functions on $\Omega \subseteq \mathbb{R}^n$. The L^p spaces, to be defined in what follows, are contained in L^0.

© The Author(s) 2015
L.T. Gomes et al., *Fuzzy Differential Equations in Various Approaches*,
SpringerBriefs in Mathematics, DOI 10.1007/978-3-319-22575-3

Definition A.4. Let Ω be a measurable set, $f \in L^0(\Omega)$ and

$$\|f\|_p = \begin{cases} \left(\int_\Omega |f(t)|^p dt \right)^{1/p}, & \text{if } 0 < p < \infty \\ \text{ess sup } |f|, & \text{if } p = \infty \end{cases}. \tag{A.1}$$

The space $L^p(\Omega)$ is the collection of the equivalence classes of all Lebesgue measurable functions such that

$$\|f\|_p < \infty$$

and equivalence means $f \sim g$ iff $f = g$ a.e.

The *ess sup* is the *essential supremum*,

$$\text{ess sup} f = \inf\{c \in \overline{\mathbb{R}} : \mu\{\omega : f(\omega) > c\} = 0\}$$

that is, the smallest value c such that $f \leq c$ a.e. Hence $\|f\|_\infty < \infty$ means that f is bounded except on a set of measure zero.

If Ω is a measurable set, we say that f is *Lebesgue integrable* if

$$\int_\Omega |f(t)| dt < \infty.$$

Definition A.5 (See, e.g., [2]). A function $f : [a, b] \rightarrow \mathbb{R}$ is called *absolutely continuous* if, given $\epsilon > 0$, there exists $\delta > 0$ such that for every countable collection of disjoint subintervals $[a_k, b_k]$ of $[a, b]$ such that

$$\sum (b_k - a_k) < \delta$$

we have

$$\sum (f(b_k) - f(a_k)) < \epsilon.$$

Equivalently, we can say that a function $f : [a, b] \rightarrow \mathbb{R}$ is absolutely continuous if it has a derivative almost everywhere (that is, except on a set of measure zero) and

$$f(x) - f(a) = \int_a^x f'(s)\, ds.$$

This result is stated in the Lebesgue Theorem:

Theorem A.1 (See e.g. [3]).

(a) Let $g \in L^1([a, b])$ and

$$f(x) = \int_a^x g(t)dt, \quad x \in [a, b].$$

Then f is absolutely continuous and f′ = g a.e.
(b) Let f : [a, b] ∈ ℝ be an absolutely continuous function. Then f′ is integrable in [a, b], that is, f′ ∈ L¹([a, b]), and

$$f(x) - f(a) = \int_a^x f'(t)\, dt, \quad x \in [a, b].$$

The set of all absolutely continuous functions $f : [a, b] \to \mathbb{R}^n$ is denoted by $\mathscr{A}C([a, b]; \mathbb{R}^n)$. The notation $\mathscr{A}C([a, b], \mathbb{R}^n)$ and $L^p([a, b]; \mathbb{R}^n)$ stand for the generalization of these spaces from codomain $\mathbb{R}$ to $\mathbb{R}^n$. Derivative and integral are calculated term-by-term on the n-dimensional vector. A subset in $\mathscr{A}C([a, b], \mathbb{R}^n)$ that will be used is

$$Z([a, b], \mathbb{R}^n) = \{x(\cdot) \in C([a, b]; \mathbb{R}^n) : x'(\cdot) \in L^\infty([0, T]; \mathbb{R}^n)\}$$

and will be denoted $Z_T(\mathbb{R}^n)$ when $[a, b] = [0, T]$.

References

1. R.B. Ash, *Measure, Integration and Functional Analysis* (Academic, New York, 1972)
2. J.P. Aubin, A. Cellina, *Differential Inclusions: Set-Valued Maps and a Viability Theory* (Springer, Berlin/Heidelberg, 1984)
3. C.S. Hönig, *A Integral de Lebesgue e suas Aplicações* (IMPA, Rio de Janeiro, 1977) [in Portuguese]
4. W. Rudin, *Functional Analysis* (McGraw-Hill, New York, 1973) [in Portuguese]

Index

© The Author(s) 2015
L.T. Gomes et al., *Fuzzy Differential Equations in Various Approaches*,
SpringerBriefs in Mathematics, DOI 10.1007/978-3-319-22575-3

Printed in the United States
By Bookmasters